TAPPING INTO THE TOTAL POWER OF THE UNIVERSE

Time and Gravity Control Technologies for the 21st Century and Beyond

PAUL W. LANDER

TAPPING INTO THE TOTAL POWER OF THE UNIVERSE
Time and Gravity Control Technologies for the 21st Century and Beyond
All Rights Reserved.
Copyright © 2019 Paul W. Lander
v5.0 r1.2

The opinions expressed in this manuscript are solely the opinions of the author and do not represent the opinions or thoughts of the publisher. The author has represented and warranted full ownership and/or legal right to publish all the materials in this book.

This book may not be reproduced, transmitted, or stored in whole or in part by any means, including graphic, electronic, or mechanical without the express written consent of the publisher except in the case of brief quotations embodied in critical articles and reviews.

Outskirts Press, Inc.
http://www.outskirtspress.com

Paperback ISBN: 978-1-9772-1422-5
Hardback ISBN: 978-1-9772-1411-9

Cover Photo © 2019 SSRO/PROMPT and NOAO/AURA/NSF. All rights reserved - used with permission.

Outskirts Press and the "OP" logo are trademarks belonging to Outskirts Press, Inc.

PRINTED IN THE UNITED STATES OF AMERICA

Acknowledgements

All research discussed in this book was a team effort. The team included several talented individuals. The author's part was to conduct the experiments and communicate the results to each team member. Through written correspondence, ideas and conclusions were drawn from a wealth of data that was gathered. A small portion of that data has been incorporated into this book. The author would like to acknowledge the following team members and their important contributions:

Carlton (Carl) Irish: Carl can be described as a "walking book of knowledge". Carl was a WW II and Korean War veteran and an engineering supervisor with Bell Laboratories, where he worked on the early development of the maser and microwave technologies. Carl was my physics professor when I was an undergraduate student. Carl completed the brilliant design for the original test apparatus. He was my mentor and friend.

Biaggio (Basil) DiPietro: Basil was introduced to me by a mutual friend. We immediately became friends, as we shared the same passion; to discover and develop a new means of ultra-rapid transportation. It was a wonderful feeling to be able to exchange ideas on how to go about achieving a major scientific breakthrough in this area. It was Basil who insisted that the EMMA craft design held the key to what he called "freedom travel".

Richard (Rich) Morongell: If you are looking for a creative or alternate means to solve a problem, Rich is the man you need on your team. Rich used both his mechanical engineering and architectural skills to make design modifications to the original experimental apparatus so that more consistent results could be achieved. He was also concerned about my safety and made changes around the moving components of the apparatus, which greatly improved its stability. Together, we worked out the details on how to extract electricity from the ground. The graphics in this book were completed by Rich.

Nicholas (Nick) Reiter: Nick was intrigued by the unknown. Nick investigated what many considered the paranormal, leaving no stone unturned. His methods of research employed a scientific approach, while reporting accurately on what was observed by a variety of observers. There was no sensationalism in his work. He accurately reported on what appeared to be unusual phenomena. He was a scientist with an open mind. Curiosity is a powerful driving force and Nick had plenty of it. Nick's knowledge of electronic circuitry was invaluable.

Jim Hopper: I got lucky when I found a machinist only a few minutes from where I performed experiments over a period of thirteen years. But Jim was not your ordinary machinist. Jim was a tool and die maker and perfected his skills as a machinist. When I met him, he had accumulated several decades of experience. His knowledge of various materials and their tolerances was incredible and invaluable. He was a perfectionist, a gentleman, a friend, and always ready to help in any way he could.

Table of Contents

Acknowledgements i

Introduction v

PART I: GRAVITATIONALIZING A MAGNETIC FIELD

1. Analysis of An Experiment 1

PART II: DEVELOPING TIME AND GRAVITY CONTROL TECHNOLOGIES

2. Creating A Time Distortion-Experiment Conducted on March 14, 1988 15
3. Converting Solid Matter to A Confined Energy Form 20
4. Superimposed Mate Pairs of Universes 26
5. The Three Phase States of Time 28
6. Magnetism, Minite Plasma Clouds and Gravity 30
7. Controlling the Lock Force 33
8. Super High Frequencies 38
9. Manufacturing A 3-D Earth 39
10. The Photon Lightron Chain Reaction Force 40
11. Straight Spurts of Interrupted Quantum 43
12. Producing A Gravitational Field 47
13. Dual Spectrums and Harmonic Amplification 49
14. Dimensional Conveyors 52
15. Interstellar Communications 54
16. Superconducting Electrons and Positrons 56
17. The EMMA Craft – Creating A Life Line to The Stars 58
18. Extracting Energy from the Earth 65

PART III: CONCLUSIONS AND FURTHER INSIGHT 69

19. Summary of Conclusions and Further Insight 71

PART IV: GLOSSARY OF TERMS 85

Introduction

In 1986, I was part of a team of investigators conducting research and experiments involving counterrotating electromagnets. The experiments were designed to see if we could find a relationship between electromagnetism and gravity. Experiments were conducted during a period of 13 years. These audible and visible results were video recorded and color photographs were subsequently produced from these recordings.

Many of these experiments produced ejections of particles or plasma into the rotating magnetic fields that captured and accumulated them. This caused the neutral rotating magnetic fields to become electrified and by expanding or condensing the electrified magnetic fields, all matter in the surroundings expanded or condensed with them. As the electrification of the rotating magnetic fields occur, the electrified magnetic force then becomes the dominant force not only over all matter, but gravity and time as well. Because the surrounding matter is expanded or condensed, time and gravity are either reduced, increased or eliminated. When gravity and time are eliminated, the entire universe opens up to our exploration.

The purpose of this book is to open a door that will lead to the development of a more advanced science and technology. The idea of time travel and space travel has been written about in science fiction novels and is now a part of our culture, as exhibited by our interest in entertainment such as *Star Trek and Star Wars*. However, the means to accomplish time travel and space travel has remained elusive. The problem is, science has never been able to define time or gravity in a way that allows us to put these two forces of nature to practical use... Until now.

Time control technology can be used to regress aging, as well as any disease. The control of gravity and time will open up a whole new era of technological discoveries that will far surpass what has been written in the most imaginative science fiction novels. In addition, unlimited clean energy can be tapped into for power generation and a new clean industrial revolution will take place. Also, almost any matter can be created from energy, including food, by condensing energy into matter using the force of gravity.

With the use of this new science and technology, immediate entry to any location on Earth or in the universe can take place. This is because once time is eliminated around a properly designed craft, all of space becomes an instant entry. In addition, one can travel by a different means called "gravity in the direction of flight". This is an ultra-fast means of mass transit or private transportation, where you can travel to any place on Earth in less than one hours' time.

Science has long demonstrated that all atoms have a nucleus core that is surrounded by separated energy levels called shells. Quantum theory states that these shells are at fixed or unchanging distances from each atom's nucleus and that in these shells reside particles we call electrons. These particles appear as cloudlike structures that revolve at enormous speeds within each shell.

These electrons move in different orbits, like the planets that orbit our Sun. However, when an atom receives energy by absorbing light, one of the electrons jumps to a higher orbit (shell), further out from the nucleus. But it does so without ever passing through the space between shells. So, when an electron absorbs a photon of light, it literally disappears from its shell or fixed orbital distance from the nucleus, and immediately reappears at a more distant fixed orbit or shell. Then the electron jumps back down to its original shell. What this means is that electrons can only absorb and emit fixed amounts of energy or photons of only specific wavelengths and frequencies, that come in certain size bundles or packets called quanta. These quanta are separated or interrupted by a tiny interval. Thus, quantum theory states that energy (light) travels in interrupted quantum packets or bundles and that electrons absorb and emit energy in steps of an interrupted quantum.

This strange behavior of electrons has been well documented. However, what puzzles many scientists is, how can the smaller world of atoms and molecules behave in this strange manner, while the larger world of people, objects, planets, stars and galaxies, behave so differently? After all, everything in the larger world is made from these same atoms and molecules and we don't see people or objects disappear and immediately reappear elsewhere. Also, at what point does the smaller world of atoms and molecules seemingly irrational behavior cease, and the larger world of people and objects "normal" behavior begin? In this book, we will explain how the larger world of people and objects, placed in a specially designed craft, can be made to disappear from one place and immediately enter another, by either expanding or contracting the so called "fixed" or unchanging energy shells of atoms of the entire craft and crew.

Part I of this book will focus on a particular experiment. Color photos of the experiment are provided. Part II will focus on the development of time and gravity control technologies and tapping into the total power of the universe. Part III will summarize the conclusions we drew from our experimental results and also discuss a few compelling topics. These topics include, "What is controlling the hologram universe we live in and how can we alter the program to serve our own needs?" Part III will also explain why it is important to begin to incorporate this more advanced science and its potential technological advances into our daily lives. The goal is to create a sustainable and much improved way of life for all humanity. Lastly, Part IV will consist of a glossary of terms.

PART I
GRAVITATIONALIZING A MAGNETIC FIELD

Analysis of An Experiment

The photograph at the top of Page 2, was taken of the original experimental apparatus assembled in May 1987. The other photograph on Page 2, shows some of the structural and component changes made to the apparatus over the period of investigation (1986-1999). Beginning with the very first experiment conducted in May 1987, unusual effects began to occur. But these odd effects occurred randomly.

So, we made careful observations until we determined their cause. At first, we eliminated the cause, but later established conditions required for reproducing the effects at will. Conditions required for reproducing these effects, allowed for the expulsion, capture and accumulation of energy releases into the rotating magnetic fields of the apparatus. Typically, when energy is released from any device or vehicle, whether it's from an automobile or plane, the energy gets dispersed into the environment. That energy is lost and becomes a waste product or pollutant. On the other hand, when energy (in this case electrons) is released, captured and accumulated, until it reaches very high proportions or intensities within the rotating magnetic fields, unusual effects begin to occur in the surrounding environment, under certain conditions.

As time went on, our team of investigators had many discussions about the experimental effects and their implications, soon referring to some of those effects as "gravitationalizing" a magnetic field. Why? While we have learned how to use magnetic fields to accelerate electrons of atoms into linear conductive paths, creating an electrical current flow, we have not yet been able to alter magnetic fields so that they additionally cause the nuclei of atoms to accelerate along with the electrons. But when a rotating magnetic field becomes "electrified", it acts upon the total makeup of all atoms and a one-way (linear) movement on all matter objects occurs, similar to an object falling in a gravitational field. This causes the magnetic field to act like gravity, keeping in mind, magnetic fields can be made much stronger than gravity. In other words, once a simple to produce magnetic field can be altered to act like gravity, which has an effect on all matter, anything or anyone can be swept along with the movement of a "gravitationalized" magnetic field. This is because the more powerful electrified magnetic force, now becomes the dominant force, over all matter and gravity, as you shall see in the following experiment.

If this "gravitationalized" magnetic field is made into a concentrated beam, any object or person can be drawn toward or projected away from the beams point of origin. In addition, these beams can be used to propel or project any size aircraft or spacecraft at unheard of speeds. This will be discussed in more detail in Part II.

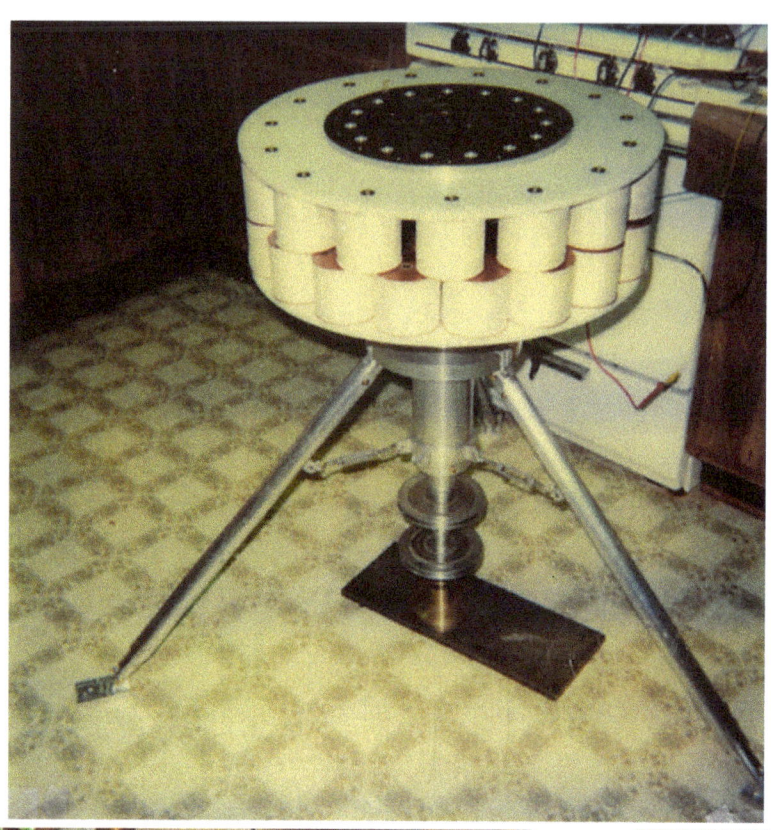

2 | TAPPING INTO THE TOTAL POWER OF THE UNIVERSE

Please see Page 4 - **Top Photo**. The photo at the top of the page depicts the apparatus prior to the start of a particular experiment, which was conducted on January 18, 1996, almost two hours after sunset. Before beginning the experiment, **all room lighting was shut off**. The electromagnets, seen covered with yellow or white reflective tape, are sandwiched between the two large aluminum plates.

The experiment began by setting the two electromagnetic coil assemblies, each having twelve electromagnets that were bolted to one-inch thick aluminum plates, in counterrotation, using two direct current motors. Soon afterward, direct current electrical power was gradually fed to the counterrotating electromagnets to produce an electrical current flow in them. Less than two minutes into the experiment, flashes of light were emitted around the apparatus. The light flashes were being emitted from matter that surrounded the apparatus, which included the garage door and the wood support structure of the apparatus. As the charged particle electrons released from the rotating apparatus began to accumulate in the immediate vicinity of the rotating field coils, the neutral magnetic field became highly electrified, causing the flashes of emitted light to become more intensified. Soon, these light flashes were accompanied by loud high-pitched sounds, identical to sounds made by a whip.

Page 4 - **Bottom Photo**. Observe how substances including wood, metals, insulated wires, and plastics, have expanded their matter makeup and emit light, in the **darkness of the garage**. Observe how all matter in the surrounding environment is being sublimated into a rotating plasma field, due to the actions of the electrified magnetic vortex. In other words, all the surrounding neutral matter has become electrified and is being attracted to and or swept along with the rotating magnetic fields. Note the 2.0" x 6.0" rear vertical pine wood structural support that has expanded and is emitting light, now appears in front of the counter rotating coil assemblies.

Atoms are all made up of equal amounts of negatively charged electrons and positively charged protons, making them neutral. But, as some electrons from the rotating apparatus were forced outward, they began to accumulate in the rotating magnetic fields. Then the surrounding neutral matter such as the wood, became de-neutralized or electrified, as the electrified magnetic rotating fields passed through them. Since charged matter is attracted to magnetic fields, all the surrounding matter that has become charged or de-neutralized, is swept along with the rotating magnetic fields.

Page 5 - **Top Photo**. Notice how the black insulated conductive wires that provide power to each electromagnet, have stretched, shifted color from black to brown, appear "taffy-like" and are now in front of the electromagnets, instead of behind them, where they are fastened by small machine bolts, to the aluminum plates. Also notice that the reflective tape around each coil is now blue, rather than yellow or white, while the aluminum plates appear stretched out, similar to the "taffy-like" wires. Page 5- **Bottom Photo**. On the left side of the photo, a small explosion took place as particles or clusters of particles collided, due to the particles counterrotational movement. One can see the radiant energy that sweeps across the apparatus, while the electromagnets appear to be slightly compressed on one side, as the radiant energy release sweeps from left to right in a slightly downward motion.

Photo Page 7 - **Top Photo**. This photo shows a rear view of the apparatus. It was taken prior to the beginning of the January 18, 1996 experiment. Notice how the 1.0" thick oak cross braces are fastened to the 2.0" x 6.0" vertical pine supports with small brass screws.

Photo Page 7 - **Bottom Photo**. This photo was recorded a fraction of a second after the photo that appears at the bottom of Page 4, where the vertical pine support expanded and was emitting light, until it snaps back to its original position. Looking at the bottom photo of Page 7, one sees the vertical pine beginning to reestablish its normal state of solidity and becoming a piece of serviceable pine wood again, while the oak cross brace on the opposite side of the apparatus, has expanded, appears to be on fire, temporarily attaches itself to an electromagnet, and has moved in front of the apparatus, as the assemblies rotate. Again, we can observe the stability of the surrounding matter being affected by the electrified magnetic fields in rotation. Somehow, the proper modulation or integration of the surrounding, rotating, electrical and magnetic forces in either a continuous compressed or expanded state, generates an invisible force we call gravity. We will go into this in more detail later on.

A fraction of a second later, the oak brace snaps back to its original position, where it was attached to the brass screw of the vertical pine support (see top photo Page 7). A loud sound is heard, as the oak brace returns to the brass screw, that is identical to the sound produced by a whip. While the oak brace appears to be on fire, what we call "solid matter" (solid, liquid or gaseous), may be nothing more than frozen light. When energy travels slower than the speed of light (approximately 186,000 miles per second), the energy must condense to form different states of matter solidity, depending on its speed over the speed of light. In this experiment, the surrounding matter expands, begins converting to energy-light, as though it were being accelerated to near light speed, while the device remains in a stationary position on the ground, rotating at a relatively slow speed. Not seen in this photo or other photos from similar experiments, were the lower 1.0" thick oak braces, which "arced" to ground, as they became electrified.

Because the charged particles moving within the magnetic vortex are not equally distributed, the matter around the apparatus is being affected with greater or lesser intensity, depending on charged particle density variations within the magnetic vortex, as the device rotates. Certain field intensities of the charged particle magnetic fields are needed, in order to affect the stability of the surrounding environment. If the apparatus was encompassed by a properly designed shell, the electrical particles would be equally distributed within the revolving magnetic vortex and the apparatus would soon convert to a virtual non-mass or energy form, along with the immediate surroundings.

Page 9- **Top Photo**. This photo shows the experiment winding down. The room lights have been turned on, the power to the electromagnets was turned off, while the coil assembly's rotation was allowed to continue. Notice that the vertical pine wood has not yet completely returned to its original state of solidity. One can still see some reddish-brown sponge-like substance just over the top aluminum disk on the right side and in front of the vertical pine. As the produced field intensity diminishes, the surrounding matter begins to stabilize or neutralize the unstable affected matter. If the surrounding affected matter can't be neutralized locally, it will be immediately dispersed away to its sister state to be neutralized there. Its sister state can be anywhere in the universe. How this process works, will be explained in greater detail later on. This is the essence of Tapping into the Total Power of the Universe.

Page 9- **Bottom Photo**. Notice the dark patch in the top coil assembly, where an electromagnet exists. The electromagnet in this position appears to be in a state where it is not absorbing and emitting visible light.

Page 10- **Photos**. These photographs are simply larger photos of ones you have already seen. They were included because they allow you to see in greater detail, what is happening to all matter surrounding the counter rotating apparatus.

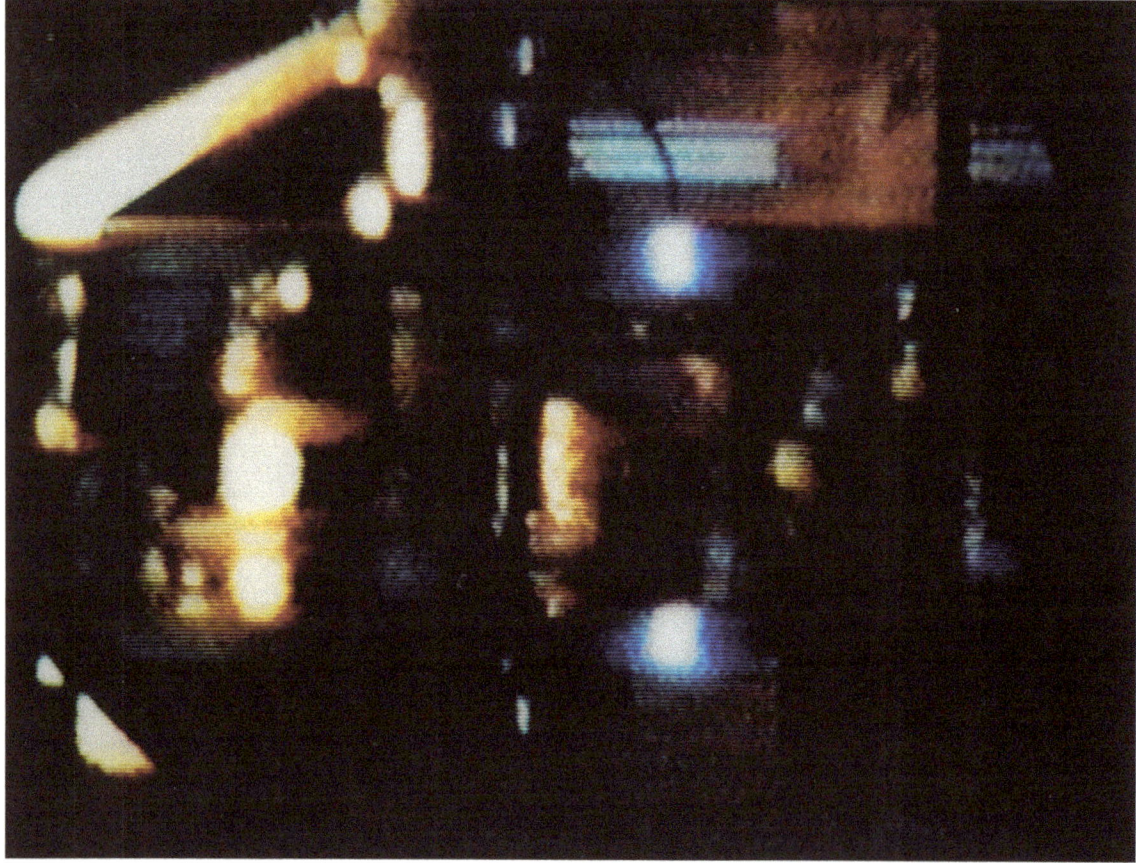

The produced synthetic revolving gravitational field or electrified magnetic field can be diverted into any linear action, to cause flight, thus producing a one-way material stress in matter. A new means of air and space travel can be developed. This means of transport causes all the atoms of the craft and crew to gain energy, rather than lose or expel energy (e.g. rocket or jet). Then, when the threshold of the atomic mass makeup is achieved (the maximum amount of energy that should be induced into any matter), the overload energy is expelled as light and sound, rather than noxious gases. The energy injection principal requires energy to be accelerated and induced intermittently into a specially designed vehicle, into the direction of travel and can be compared to pushing a swing at the right moment.

The energy injection principal can also be compared to throwing a ball. Energy is induced from the person throwing the ball into the ball's atoms, causing the ball to accelerate in the direction of the applied action. This causes an energy gain to each of the balls atom's electrons. This energy gain then causes the balls electrons to move in ellipses about their atomic nuclei, thus creating an atomic strain or stress in the direction of the ball's flight path, as the negatively charged electrons, produce numerous tiny tugs on the balls positively charged nuclei.

By changing the orbital velocity of the electrons around the nucleus, the so-called "slingshot effect" changes the rotary motion into one-way linear motion. But due to outside forces, such as gravity, its energy becomes stabilized and is subject to the outside force, because the ball has lost its stress field energy and falls to the ground. To keep the ball moving in its flight path, periodic timed energy inducements are required. Perhaps this can be accomplished by somehow catching up to the ball while in flight and throwing it again, just before the ball begins to follow a curved downward path toward the earth's surface or by inventing a means of energy inducement, using a device that is placed in and affixed to the ball. It is the same device that can be placed in a craft, to cause linear acceleration in any direction.

To give yet another example of the principle, picture if you will, two electrons orbiting the nucleus, one in one direction and the other oppositely. Now, at a certain point around their orbital path, when they are 180 degrees apart, a partial blocking force is applied to each electron. Since these electrons have gained rotational energy in their path, this energy is given off and is attracted to the nucleus, creating a locked fixed condition. Since the nucleus and electrons are locked together, the entire atom has no alternative but to move in the applied blocking force direction.

The nucleus then receives the energy to the point of its threshold, then gives off its energy to the electrons and the process is repeated, as the electrons reach their threshold and release energy back to the nucleus. If the electron elliptical path is overextended, the threshold is exceeded. Then the electrons escape the nucleus and energy is given off in the form of sound and light, plus other energies that we cannot detect by any means. These energies are in a range that is unknown to us. While they can create beneficial conditions of time and space distortions, surpassing the threshold has certain limits. If those limits are exceeded, deterioration of the crafts matter will begin to occur. One must calculate the total mass makeup of the craft and crew in order to determine the threshold of injected energy.

This is an example of how a new type of quiet, ultra-fast, non-polluting transportation functions.

We call this type of craft EMMA. It will be discussed in greater detail in this book. Drawings of the EMMA crafts linear drive system has been provided. This means of transportation will replace all forms of transportation we currently have. It will not only be used around our planet, but throughout our solar system and beyond.

PART II

DEVELOPING TIME AND GRAVITY CONTROL TECHNOLOGIES

2

Creating A Time Distortion-Experiment Conducted on March 14, 1988

The photo on Page 16 depicts what appears to be changes to space, time and matter in the immediate area of the experimental apparatus. The arcing around the electromagnets occurs in free space. The arcs did not discharge to ground, either through the aluminum plates, which were not grounded or through any of the electromagnets, which would have been destroyed. Less than 275vdc were being applied to the electromagnets, with each of the 24 electromagnets, connected in parallel, receiving about 10.0 milliamps or 0.01 amps of power.

Concurrent with the arcing in free space, the coil assemblies began to accelerate on their own (without increasing the motors speed), soon reaching speeds greater than 500 rpm's. These speeds were well beyond what we calculated the 5/16" thick brass bolts, used to fasten each electromagnet to their aluminum plate, can tolerate. Even after shutting off the two DC motors used to counter rotate the two coil assemblies, the electromagnetic coil assemblies continued their acceleration. It was only after power to the electromagnets was also turned off, and an additional 20 seconds of continued acceleration took place, that the coil assemblies finally began to slow down.

Afterward, the electromagnets and electrical connections were carefully examined. It was found that none of the brass bolts had bent, and none of the electromagnets experienced any deformation or damage of the slightest kind. Not even a mild scraping of the 1/16" thick phenolic plastic flanges occurred. Gaps between top and bottom coil plastic flanges, at the center point of the assemblies, were kept at slightly more than 1/16" inch or about 70 thousandths of an inch. These flanges should have been torn off and the copper magnet wires within these coils, ejected outward into the garage.

Note that some of the electromagnets appear (see following page) to have moved inward from the periphery of the aluminum plates where they are bolted. These electromagnets should have tilted outward, if they were being affected by centrifugal force. Note that rotation is a form of acceleration, because a constant change in direction occurs, due to circular motion. This is called centripetal acceleration. However, an additional form of acceleration was somehow created, causing the coil assemblies to continuosly increase in speed, even after the motors and power to the electromagnets were shut off.

Also, in the photo on Page 16, note the two "emanations" at the top left side between the aluminum plates. Observe the emanation on the left with ball-shaped head, necking down to what appears to be an upper and lower body, with perhaps two legs. Emanations occurred in several subsequent experiments, along with bursts of plasma ejections around the apparatus. Considerable time was spent interpreting the experimental effects that took place over a thirteen-year period and their implications.

This is part of what we concluded: As the coil assemblies counterrotate, they cause a continuous accumulation of ejected charged particles, along with the stretching or **breaking** of the magnetic field lines of force, plus the unification of the electric fields and magnetic lines of force into a single field of force. The mechanical unification of these two fields, means they are no longer traveling at right angles to each other and as their kinetic forces combine, they can have an effect on the surrounding matter atoms and molecules, under certain conditions.

Furthermore, if time is somehow being reduced around the apparatus, then distance had to be reduced too. If distance is being shortened, the coil assemblies had to accelerate, as the coil assemblies would then have less distance to travel in their rotation. The shortening of distance around the apparatus would also explain why some electromagnetic coils appear inward from the periphery of the aluminum plates. Since time and gravity are related (Einstein's Theory of General Relativity), gravity is also being affected. But what happens to the speed of light (approximately 186,000 miles per second) measured in Earth seconds, if time (seconds) is being reduced?

Does the speed of light change or is there is an illusion of the light speed barrier being altered, created by a time distortion? Or is there a distortion of matter, which appears to be what is occurring? Perhaps time is nothing more than changes in matter, as we witness changes taking place around us. Currently, we believe that atoms do not change as time passes, but then how can we explain why we can't see past or future events, if all events that occur around us, are made up of atoms that don't change (expand or contract)?

After years of experimentation, we recognized the need to define gravity and time in a way that helped to explain the experimental results obtained. This is what we concluded: As Earth and other space bodies travel through space, they must lose energy. Nothing can move without a loss of energy. As the Earth spins on its axis and orbits the Sun, as well as the galaxy, energy that is being continuously lost, must be expelled from every Earth atom associated with these motions through space. The larger a space body, the faster it travels, the more energy it loses. Now, if space were permeated with a highly expanded, undetectable energy, then this space time energy could act as a cosmic equilibrium force, becoming attracted to condensed matter energy distribution points, such as planets and stars.

Time and The Universal Equivalence

The cosmic equilibrium force or universal equivalence is what we call "gravity", while the energy loss from space time bodies' (e.g. planets and stars) atoms is what we call "time". As time loss energy releases occur, along with the changing of matter energy atoms, they transition to form a shadow universe called the past, as they become imperceptible to our senses and undetectable to our detection equipment atoms, that are also changing at the same rate as their surroundings. We can't see or detect past or future events, because all atoms are in a constant state of gradual expansion, with changing energy loss rates that must be unique for each universal space body, such as our planet Earth.

Thus, time energy expelled from matter becomes a negative factor, causing an attractive force or universal equivalence in the form of gravity equalization, derived from a cosmic equilibrium force that permeates all of space. The atoms we are familiar with are of a certain peripheral expansion now, but as what we call time proceeds, these same atoms are expanding into a more expanded existence and expel more energy, that becomes the new now. Since energy cannot be destroyed, it must be changed to different spaced gradients. This means that the shells within atoms that contain particles such as electrons, do not remain at fixed distances from each atom's nucleus, as quantum theory states. The shells containing electrons gradually expand outward from the nucleus, increase in orbital speed, and expel more energy. To help you to comprehend what's occurring, as the shells or energy levels expand outward to cause the illusion we call time, picture a constant changing between each ripple spacing on a pond after a pebble hits its surface. We can only see the present ripple spacing, as our energy makeup of atoms changes (expands) at the same rate as our surroundings. We can't see past or future atoms state of solidity.

An atoms consistency is relative to its position in the universe and relative to its speed over the speed of light. Each space-time body or planet for example, has a unique meaning for the term "solidity" and a unique range of consistency for its atoms. What we call solid matter on Earth, would not be considered solid anywhere else. What we called solid yesterday, would not be considered solid today. Tomorrow's state of solidity for atoms is not the same as todays. Tomorrow's atoms and events are in a more expanded state and yesterday's atoms and events are in a more condensed state and we cannot see or detect them, unless some local anomaly occurs or a new science and technology is developed. We are a part of a single, elastic, pliable existence called the universe that is stretching.

All waves, particles and objects traveling through space, toward a planet such as ours, would experience gradual changes along their travel path through space and when they reach the Earth, the waves of light for example, can then be absorbed by the Earth's atoms current state of atomic solidity. The particles and objects upon arrival would appear solid and real to us, because now their energy consistency matches that of Earth's. If you walk across a room, all the atoms of your body rearrange themselves internally with each step you take, to match each point of entry. If you didn't change, you couldn't move. But these changes are very minute, because you haven't gone very far.

Universal Atomic Coding and Universal Time Constant

When light is emitted from a star, it represents that stars unique atomic coding or particle orbital spacing within atoms. However, in order for it to enter another position in space, the emitted light must change to that position's accelerated fixed point. Therefore, light changes along its path with each new point it enters. When it reaches Earth, it has matched Earth's space time body's coded spectrum sequence or particle orbital spacing within atoms, and is detected, meaning it is absorbed by Earth's atoms. If light didn't change along its path, there would be no different space-time energy points to enter.

Picture yourself running around a curved race track. As you move faster, you tend to move toward the outside of the track. Just like electrons traveling around the nuclei of atoms, you are moving faster and further out from the center, while expelling more energy. Like ripples in a pond after a stone is tossed in, the electrons and nucleus particles form more outward orbital paths with the passage of time. Therefore, there is a harmonious relationship between atoms and the celestial bodies they make up, as they accelerate through space together. All our surroundings are in a constant state of change. We have called this change 'time'. Trapped here on Earth, we are all moving together at the same acceleration rate through space.

Yesterday becomes an invisible and imperceptible world of atomic and molecular makeup. No two atoms can occupy the same space at the same time and no two atoms in the universe would have the same orbiting speed and orbit for their particle makeup. Each atom has its own coded unique energy consistency or spacing between particle orbits. What we think of as solid matter, should be thought of as primarily empty space, along with different energy mass bandwidths in a state of change, due to changes in speed. In the case of the formula $E=mc^2$, the ratio of mass and the speed of light experience changes in the mass energy bandwidth. This means that we need to begin to understand the relationship between mass and its speed or acceleration and the different energy mass bandwidths that exist throughout the universe.

One must recognize that matter energy forces are directly related to their speed in comparison to the constant speed of light. By expanding or contracting matter at the atomic level, the light spectrum is shifted, causing a condition of time warping, as unknown to Earth quantum packets are being released. For example, the visual light waves change and if you try to observe them, you can't. They are now beyond the range of detection or observation, but they still exist. They have become stretched into another dimension and it is this stretching action that will cause speeds faster than light because the universal time constant as we know it, has changed. It is these unknown to Earth quantum packet releases, that can change the surrounding atoms state of atomic solidity, by assisting in the adjustment of particle orbital speeds and distances from the nuclei of atoms.

Picture each atom as a miniscule vortex of condensed energy. Now picture creating a massive and intense electrical and magnetic vortex that can be expanded or contracted. By expanding or contracting the larger more powerful energy vortex, the much smaller atoms that makeup the surrounding matter within the larger energy vortex expand or contract too, as they become overwhelmed by the more powerful energy vortex, creating what can be called a "time distortion". This is because the so called "fixed energy levels" or particle orbits within atoms have changed (expanded or contracted). This means that by slightly expanding or contracting matter atoms, from their nucleus outward, time can be reduced, increased or eliminated. Eliminate time and gravity is eliminated. Chapters Seven through Eleven will discuss how matter atoms and molecules can be altered, so that time and gravity can be affected to achieve a desired result.

3
Converting Solid Matter to A Confined Energy Form

A space craft and crew, traveling to the Moon, will experience slow quantum (tiny) changes to all its atoms along its travel path, until they eventually reach their objective, a particular location on the Moon's surface. A tremendous amount of energy is expended, just to be free from the Earth's gravitational field. On the other hand, if a space craft were designed to make predetermined quantum changes in the beginning while still on Earth, to its total atomic makeup, to match a specific location on the Moon's surface, it will be transmittedly changed for immediate entry to its matched condition. How will this occur?

The entire craft and crew initially appear as a plasmoid form, similar to what can be seen in the photos in Chapter One, followed by a further stretching, until the matter becomes a virtual non-mass state or confined energy form. This causes the craft and crew to become invisible and merge with equivalence flow (energy traveling at the speed of light), like a drop of water that is placed in a flowing river. Once this occurs, one can visualize the universe below the craft as a giant football stadium with numerous seats, each seat representing a galaxy or solar system. Now, upon entry to the equivalence flow, the craft automatically deploys its preprogrammed computer controls to establish conditions for taking up a seat, by creating a particular universal coordinate state of atomic solidity, which in this instance is the Moon. In other words, the craft's matter is first stretched into a confined energy form and then compressed in order to enter whatever matter state of atomic solidity desired, such as the Moon.

When the desired matter atomic consistency change is applied to the craft and crew, while in the equivalence flow, the universe's universal equivalence recognizes the unbalanced condition that has been created and employs its total power, to shift the vehicle and crew to the proper location within itself, the Moon, where neutralization of the craft's atoms can best take place. The universe does this by opening a "void hole", where no gravity or time energy exists, which assures zero resistance to motion. Then, it gently pushes the vehicle through the "void hole" to its matched condition, the Moon.

Each space time body has its own unique resonant energy bandwidth and matching gravitational field. Duplicate a particular planet or moons resonant energy bandwidth around a space craft and immediate entry becomes a reality. This occurs because the universal equivalence of space time energy becomes attracted to the craft in the form of a planet's or moon's gravitational fall rate or time continuum field of existence. So, by duplicating our Moon's gravitational fall rate around the craft, the craft immediately enters the Moon's environment. The space craft is "transmittedly" changed to immediately enter its matched condition. When the craft enters the equivalence flow, it becomes non-conformant to all known universal laws and can make up its own laws, as desired by the crew.

Call this freedom travel. Picture our universe as a vast river of energy and most of that energy is invisible and flowing at the constant speed of light or approximately 186,000 miles per second. Some of that energy, a smaller portion, has condensed to form planets, stars and galaxies. If you look closely at a flowing river, you may see different size spirals of water moving along with the flow. The spirals of water will move at different speeds, but none will move at the speed of the river itself. They will always move slower.

If you take different density pieces of wood using logs of the same size and place them in the river flow, you will observe something similar occurring. The greater density wood will move more slowly, while the less dense wood will move more quickly. If you want a piece of wood to acquire the speed of the river flow, you have to change its density or reduce its mass to match that of the river. If you place a drop of water in a river the water will immediately acquire the speed of the river, because it has acquired the river's energy. In the case of the logs (or rocket ships), the river is acting upon them, so that the logs (rockets) are affected by the river and requires energy to maneuver in the rivers matter energy flow.

Like the drop of water that acquires the energy of the river, one needs to learn how to acquire a tap into the universal matter energy flow, by altering the density or consistency of matter, thereby reducing its mass to a confined energy form, so that it acts as a non-mass. The formula now becomes $E = c^2$, as m or mass has been reduced to zero. When this occurs, it's like acquiring a pass to a football stadium and being able to take up any seat you want. Only in this case, the seats are star systems. The craft becomes a part of the universal matter energy river flow and is no longer affected by planetary or stellar bodies gravitational fields. It becomes equivalence free, surrounded by an anti-mass force and since time is eliminated, travel to distant star systems appear as instantaneous arrivals.

Since each coordinate states atoms are neutral, a strong magnetic, electron or ion field combined is needed to de-neutralize them. As this de-neutralization of the crafts atoms occur, the existing surrounding matter endeavors to neutralize it. If it can't be neutralized locally, it will be dispersed away to its sister state to be neutralized there. Its sister state can be anywhere in the universe. In this way one can enter any no-time energy field, because as immediate entrance occurs, no time is the result.

We exist in a certain surrounding energy concentration for atoms and if that concentration is changed gradient-wise (expanded or condensed), we are immediately attracted to the matched atoms energy concentration, regardless of distance. All this occurs because electrons travel differently speed-wise in different space coordinates. For example, each planet has a certain standard for electron speed in an electrical conductor, which produces a standard spaced gradient surrounding magnetic field. The standard electron speed in a conductor or electron orbital speed within atoms, for any particular planet, represents that planets time continuum field of existence, or address in the universe. Time, location and atomic solidity are related. Alter the electrons "normal" speed in a conductor on Earth, to match the electrons "normal" speed in a conductor on another planet and immediate entry occurs. Entry into another planets past or future is also possible. It all depends on where and when you want to arrive. Similarly, you can return to Earth at any period in time and location you desire.

Then by changing the electron speed in the electromagnets within a craft, the craft is attracted to the time continuum adjustment and matter energy consistency of atoms applied. The key is to produce a fixed electron shell expansion or contraction to the surrounding atomic structures that matches the planet's atoms' energy consistency one desires to enter. The whole concept is built around the control of the electron speed in a system of electromagnets within a craft that can alter the time continuum. Stopping the time continuum, stops the force of gravity and by certain adjustments, a combination time continuum and matter energy atomic consistency is produced, creating a "porthole of attraction."

In the equation $E=mc^2$, c^2 is the speed of light in a second of Earth time, but time varies and so must the formula, depending on which time dimension or space time body one exists in. For example, time slows down for planets that are orbiting at a greater speed in the galaxy, therefore, light speed changes because time per second changes. How much is a second of time in a dimension (planetary environment) that is moving only 90% of time as we know it? In other words, light varies with different dimensional seconds that exist throughout space. In addition, the mass of one gram is relative only to Earth's gravity gram mass or time continuum mass energy bandwidth. These estimates are not accurate when one approaches another star system.

Although the speed of light is a universal constant, in different rates of speed in space, it acts differently. As an example, if time decreases at close to the speed of light, what happens to the speed of light that travels at 186,000 miles per second? When the speed of light is achieved, there is no time or seconds (186,000 miles per?) and what may be called a perfect equilibrium condition exists. This is because all energy losses from matter come to a halt, so there is no longer a need to replenish one's energy. You now exist in an in-between world or realm where there is no time. Can this be called the "world of the spiritual"?

Every scientist realizes the amount of energy required to travel to our nearest stellar neighbor in the manner we are now using. However, in order to reach the stars, we have to change our way of thinking to that of reducing a craft's mass, rather than adding more energy (mass) in the way of some type of fuel. By reducing the mass to virtually zero, the craft and crew become a non-mass. The craft becomes part of the Universal Matter Energy Flow (UMEF), and by the proper inducements, is changed to whatever matter energy atomic solidity desired for immediate entry and back into the "slow world of time." Back into the slow world of time means back to the condensed worlds of planets and stars, which all travel at some different fraction of 186,00 miles per second.

But once time is eliminated, the words travel, speed and distance lose their meaning. Distance becomes a zero-time energy factor because distance is only a fourth-dimension time attribute. While time is an illusion or hypothetical field of reality, created by changing atomic consistency's and a loss or gain of energy, that can be controlled. Chapter 17- The Emma Craft-Creating A Lifeline to The Stars, will discuss this in greater detail.

Time Travel

Time travel is possible. If one enters a time warp and goes back in time, his or her matter energy body returns back to that period. All your thoughts and memories remain in the skip time, so it is almost impossible to bring back ideas or knowledge of events from the future. If one enters the future, they are unaware they came from the past, because all their memories are stored from the past through the skip time to the future.

Time can be explained as a form of energy. If one goes ahead in time, one losses energy. If one goes behind in time, one gains energy. Time needs to be included in certain formulas as a form of energy. This is because as time increases, matter energy expands and therefore expels more energy, as particles increase in orbital speed. However, if time is regressed, matter energy becomes more condensed and expels less energy, as particles decrease in orbital speed. Therefore, what we have here is a universal atomic code condition, as each point in space is accelerating at a different rate, while time is relative to each time continuum's energy release movement.

Space Time Velocity Factor and Time Travel

In regard to traveling into the past or future, one must consider the space time velocity factor. First, consider the question is a glass of water half full or half empty? It all depends on if the time velocity factor of the glass was in the process of being emptied or filled. If it is being emptied, it is half empty and if it is being filled, it is half full. The process of the time velocity factor needs to be considered. So, if a person wishes to travel forward or backward in time, they recognized that all matter has the ability to be compressed or expanded, as seen in the photographs contained in Chapters One, Two and Seven.

In order to describe the present inability to properly denote time, speed and distance, the following example is provided. A tunnel 10 miles long is moving from point "B" of its forward edge to point "C" that is 10 miles away. A car in the beginning of the tunnel is at point "A", traveling at 10 miles per hour. The car reaches point "B" of the tunnel in one hour and at the same time reaches point "C". One would think that in regard to the tunnel, they traveled at ten miles an hour, but how can this be if the car is now 20 miles from point "A?" So how do you determine time, speed and distance? Only in regard to adjacent solids.

But suppose these so-called "solid objects" such as the tunnel, can also be made to expand or contract, while the car is traveling through the moving tunnel at 10 mph. Does this mean that time can be lengthened or shortened? When distance is reduced or increased, time must also be reduced or increased. So, now one recognizes that by expanding or compressing matter in any area, time and distance can be reduced, increased and even eliminated. Eliminated because once matter has been stretched into a **confined** energy form or non-mass state and merges with the universal equivalence flow or UMEF, the universal laws of gravity, time and distance no longer apply. In fact, the only laws that apply for this type of craft at that point, are the ones chosen by its crew members.

As Above So Below as Within So Without

If a nucleus that is minuscule, was enlarged to the size of a marble, its nearest electron energy level would be about 1/4 mile (1,320ft) away. The atoms we are made from and all things we are familiar with, are composed almost entirely of empty space. What does that imply? It implies that there is plenty of space inside each atom and beyond their outer most electron shell for other levels of atomic existence, that can be entered, whether on Earth or a distant space time body. Call it "inner dimensional travel" because when particle orbital distances within atoms for an entire craft and crew are changed (expanded or contracted), to match a desired location or time period, instant entry is the result. In other words, you can enter any distant place, by making internal adjustments to all your atoms. You no longer need to travel to go somewhere! You simply disappear from one place and immediately enter another.

Eastern philosophy states not to search outside yourself for answers, but instead to look within. All knowledge is within. Everything that appears external is internal. As above so below as within so without. It is the essence of the enlightened mind, which is present inside every atom of nature. Note that Buddhist teachings regard physical universes as hologram projections from a mind continuum. In other words, what we call physical and real is simply an act of consciousness, projected from a non-physical realm or mental continuum, where there is no time.

Since energy cannot be created or destroyed it must be changed into different invisible spectrums. It must go somewhere. So, if the past as we know it is atoms and energy in a more condensed state, the future must be atoms and energy in a more expanded state. Could it be that matter as we know it is being stretched into spectrums of energy light we can no longer see or detect with the instrument atoms of our "now time"? As the universe expands, the time code matter energy expands and in order for a space craft to reach or be paralleled to the expanded code matter energy existence, they must match the same atomic code. If some individual walks to reach a certain destination, they expend energy, but gain time. The time it takes to get to that destination, restores the matter energy lost, in a form of another atomic code.

With *Explorer 1 and 2*, as time continues, increases in speed occurs and as they increase in speed, they expel more energy and the energy quantum packet changes. The lost energy is a transfer of the matter makeup and quantum packet change. In other words, by changing the quantum packets, energy can either be emitted or absorbed. It is this quantum packet equilibrium that can cause a space craft to appear in any place in the universe, at whatever time period one desires. One only has to duplicate entrance modes to whatever coordinate decided upon. Rotating plasma fields of highly concentrated electrons, within a magnetic field, can affect electron shells of nearby atoms, by changing the particles orbital speeds, thereby changing the quantum packets being released and producing whatever entrance mode desired.

To simplify this statement, when one changes the atomic particles orbital speed, positions change within the atomic structure, causing a different consistency of matter energy. This change causes the matter energy to immediately enter its own kind, because matter energy cannot be destroyed, only changed. If you release air bubbles from the ocean floor, the bubbles will immediately travel upward through the water column until they reach and merge with their own kind, the atmosphere. Something similar happens to a craft and crew, when you alter their state of atomic solidity to match the atomic coding of distant worlds.

Perhaps one can travel to nearby stars utilizing a matter-antimatter propulsion system, but this will take a long-time loss energy to reach them. Each day Earth makes one axial rotation, while it is in orbit around the sun and at the same time is moving in an outer galactic orbit. In order for an inner galaxy space body civilization to reach our planet, a space craft must increase its time loss energy to match our outer galactic atomic coding because the Earth travels at a greater speed than the galaxy's inner body's, as the galaxies space-time bodies move like the spokes on a wheel. When traveling to other galaxies, galactic movement in space must also be evaluated in determining the correct atomic coding.

In accordance with this new theory, each space time point has a certain time continuum field, along with a certain time constant threshold "c" (value for the speed of light) for any matter that exists within the field. This theory then goes on to state that any time continuum field of existence for any matter can be changed, by either accelerating or decelerating subatomic particles in their orbits within atoms, to change their orbital speeds and orbital distances.

This will allow any craft or "space conveyor" to enter any time constant threshold in the universe or universes. This is because all the matter energy in each universe will be at our disposal. No matter where we go or enter, as we change each time constant threshold "c", due changes in the time continuum field, all matter energy there will assist us, as it attempts to neutralize our different "c" constant. We will have harnessed a method that will create **great energy to "ultra-dimensional" proportions**, as we learn to make changes to atomic particle orbital speeds.

4

Superimposed Mate Pairs of Universes

We live in a forward time continuum universe so that all our memories are of past events. If we lived in a reverse time continuum universe, all our memories would be of future events. In the reverse time continuum universe, the effect comes first, followed by the cause. Where there is no time, simultaneity of cause and effect prevails. But there is no such thing as time, only energy matter makeup changes.

If a reverse time continuum universe were superimposed on our universe galaxies and had a different constant for the speed of light, this universe would be undetectable to our senses and detection equipment atoms. Because of its different light speed constant, the reverse time continuum universe would have a different range for atomic consistencies and a different spectrum (range of wavelengths and frequencies) for light. For example, we know that wavelength times frequency equals the speed of light and that our atoms can absorb only certain energy emissions (wavelengths), based on our particular light speed constant. However, if another universe that was superimposed on ours, had a different light speed constant that was faster for example, its stars energy emissions (wavelengths) could not be absorbed (detected) by our universe's atoms. These energy emissions are part of a different "magnetic spectrum dimension", with a different range of wavelengths (and frequencies), and you can't see or detect what you don't live in.

Perhaps several mate pairs of universes (matter and antimatter) could be superimposed on ours. While our universe atoms cannot absorb their stars' energy emissions, their gravitational effect on our matter galaxies cannot be hidden. Each of these universes can be thought of as a single quantum energy force, operating in pairs, while the galaxies of each rotating pair, change their shells to a massive rotating central nucleus, similar to how electrons change their shells around the nucleus of an atom. As above so below as within so without.

One might visit a planet in our solar system or another solar system that is totally devoid of life, but in another atomic consistency range, it is filled with life forms. In addition, living ecosystems could exist side by side, each unaware of the others existence, while experiencing their own version of being solid and real. This could be true for our own planet. Scientists have known for a long time, that our universe behaves in a way that suggests the presence of a lot more mass then it apparently contains. In the past, they detected powerful islands of gravity, "visually" between Earth and certain distant quasars, which acted as giant lenses, splitting the quasars light emissions into two identical images. These scientists were puzzled because how could gravity sources of this size, closer than a quasar, be completely non-luminous, unless of course these islands of gravity were huge universes on another invisible plane of existence.

Perhaps another more powerful universe to our own, existing on the other side of the speed of light, was confirmed absolutely by the detection and measurement of gravitational intensities, where no seen matter exists. Perhaps these experiments conducted in Earth orbit confirmed not only distant other-dimensional physical and non-physical universes, but parallel unseen matter in and around the Earth itself. What this means is universe within universe, planet within planet, place within place and people within people.

This concept can also be understood by modifying the Superstring theory to state: subatomic particles don't exist as individual particles, but are really part of an extended object, which can be called the "superstring." We don't see the whole object, because we are only tuned into one segment at a time, the particles that make up our own time constant universe. Other parts of the object are made up of unknown time constant universes, where the subatomic particles are vibrating (orbiting) at much higher frequencies than our own and are invisible to us. In other words, we only see, hear and feel a fraction of all that exists around us.

Nonetheless, these subatomic string segments can be excited into other states, just as a string instrument can be vibrated in a series of harmonics. By producing first, second, third and higher stronger harmonics of a fixed frequency, that surpasses our matter world, enclosed in a magnetic plasma bubble or synthetic gravitational field, then perhaps the puzzle of these unseen gravitational fields or worlds will be solved.

Universes could be separated and/or neutralized by neutralizing buffer zone forces, such as magnetic fields, that operate on the molecular and atomic level and gravitational fields that act as an outer controlling agent, keeping matter confined. As gravity contracts matter energy and confines it, the magnetic fields within matter, act as the neutralizer to sustain a certain order. So, to escape this controlling order, one has to duplicate gravity and use these gravitational fields to their advantage. The key that unlocks the door to creating gravity, has to be the fact that magnetic fields are used to control the even acceleration of electrons in a conductor. Therefore, by learning to use the magnetic neutralizer to our advantage, to accelerate or decelerate the electrons in a conductor to form gravity engulfed fields, we can be free to escape to wherever in the cosmos we choose.

The neutralizing buffer zones between universes would comprise a zero-time existence, while the matter and antimatter form forward time and reverse time existences. These buffer zones would act as barrier restrictors, preventing matter from crossing the light speed barrier. But a way exists to manipulate these restrictive energies, so they act as a matter booster rather than a matter restrictor. This will be discussed in Chapter Six.

5

The Three Phase States of Time

Therefore, time has three phase states. This must be since every action has an equal and opposite reaction. The action of normal advancing time could not advance if reverse time that advanced oppositely did not friction with it, to release energy for both to proceed oppositely. All general rules should be applied to all portions of science, including time. It is the released energy that joins to form the neutralizing buffer zone force or equivalence.

Matter particles and atoms of each universe must be composed of three types of minute or "minite" plasma shells, which represent three phase states of time. While one type rotates clockwise, another rotates counterclockwise and a third type does not spin, but acts as a neutralizing buffer zone force. While our universe has a matter makeup biased configuration, our universe twin, has an antimatter makeup biased configuration. Each plasma shell makeup of electrons and positrons within the nucleus is spaced apart and revolves at right angles to each other. It is a particles total entrance density, that determines whether it exists in a matter or an antimatter state of existence. This occurs because the magnetic force acts as a bounding force around all particles and maintains a certain consistency of matter energy for each so called "particle" in each dimension, which depends on the consistency or intensity of the magnetic bounding force.

So, even though an electron and positron are of equal mass, they have different energy consistencies. The positron exists in a less dense state than an electron. Totally, these particles called electrons and positrons make up the nucleus, along with the strong force glue. Their outermost shell makeup determines if a particle, for example, is a proton or antiproton. If the outermost shell is an electron, the particle is a proton.

So, electrons and positrons makeup the nucleus particles. When they are in unison, there is stability. When forced to collide, energy is given off in the form of smaller "quantum vibrates." As the electron emits quantum spurts in one direction, the positron emits them oppositely, as these are mutual quantum spurts of energy that appear as both a quantum (particle) and a wave. So, each universe's matter is a composite of matter and antimatter. The duplex waves that are released by electrons and positrons in the nucleus combine to form a neutralizing buffer zone force called the strong force, that maintains stability. Gravity, magnetism and the strong force may be different spaced gradients of the same glue force we call magnetism.

Particles then, are nothing more than energy in motion or vortex cells of energy, whether stable or unstable. Hence, there is no such thing as particles, only a different state of minite plasma cloud actions that form different plasma cloud consistencies, that makeup what we call electrons and positrons. We are living in a state of holographic consistency neutralized plasma fields. It is better to adopt a concept of minute plasma clouds or shells that can be controlled and to forget the concept of solid particles.

One has to ask themselves, where does released energy go, where is it transferred to and what dimension is it in, when one moves or a certain article moves and energy gets released? What form did it take? There has to be different spaced gradient forms of past energy release that occupies space, that present spaced gradient energy matter cannot see, yet it is there. Now, one can see by changing the present gradient energy form of surrounding atoms, one can change universal positions or by weakening or strengthening certain energy gradient forms, different time periods can be matched and entered or viewed. Today's thoughts are yesterday's memories, but how can one think about yesterday if it's gone? Perhaps it's not gone, but is only a different spaced gradient of energy in the universal energy gradient that can be visualized and even reached, if the proper spaced gradient is applied.

6

Magnetism, Minite Plasma Clouds and Gravity

We began to work on a means of how to alter the accumulated energies around our experimental apparatus. We found that we could manipulate magnetic vortex fields to create conditions of magnetic field expansion (fission) or magnetic field compression (fusion), to partially break down the barrier of the neutralizing magnetic field, which acts as a watchman between dimensions. Experiments indicate, certain surrounding intensities of expanding or contracting electrified magnetic vortex fields are required to cause a slight expansion or compression of atoms at their base (nucleus), that can be locked in magnetically to produce different visual consistencies or location changes, by applying greater field intensities. Photo Page 16 is what we call an "alternate vibrational mode-locked visual consistency." The Earth's matter maintains a certain locked spectrum consistency of shells within atoms where particles reside, which can be expanded or compressed to enter or view another locked spectrum environment, by changing the devices surrounding spaced gradient electrified magnetic field.

What is this unseen mysterious force we call magnetic? It is a source of stored energy and the same invisible force we call gravity, except it is in matter, in motion. There appears to be two stress fields that attain an orbital path through the length of the bar magnet and extend outward to a distance in greater elliptical paths, along the bar magnets sides. The stress fields appear to enter together at one end of the magnet and exit at the other, traveling in the same direction, but upon their exit, they separate, one traveling clockwise and the other counterclockwise, creating an extended field around the bar magnet. Since these two stress fields maintain an equal outward force, along the center line of the magnet and perpendicular to its poles, the bar magnet does not move. If you place iron filings around a bar magnet, you will observe the invisible stress fields configuration.

In the case of a conductive wire, when electrical currents pass through, the electrons from a power source act as a catalyst accelerating and displacing electrons from the conductor's material, forcing a stress field or fields into a large orbit around the conductor, thereby creating what we call a magnetic field. In other words, the incoming electrons repel and accelerate electrons of the conductors matter outward into much larger orbits, causing a dispersion of the electron's energy. This force field now attracts only certain elements, such as iron, because as the "magnetic electrons" achieve an outward orbit, they are searching for atoms to accept them and only certain elements such as iron, have the orbiting space for these outward electrons that are traveling a different orbital path.

So, the magnetic field is nothing more than dispersed expanded electrons, made up of minute plasma shells or "minites", that rotate clockwise and counterclockwise, orbiting in a designed particular fashion around a conductive wire. But some of these plasma shells do not rotate. These minites form a neutral zone. Now, according to Einstein's theory, as one approaches the speed of light, infinity of the particle base expands into smaller particles to a point of infinity. So, what we are feeling in the magnetic lines of force, is this same type of expansional release. Also, it is only reasonable to assume that if a magnetic field exists in the matter dimension, an anti-magnetic field must exist in the anti-matter dimension.

Now, to increase this magnetic attractive force to include all elements, as in a gravitational field, an increase in the acceleration of the current application is needed. This causes a greater shell dispersion, meaning the shells attain a wider orbit and are further apart. So, the dispersion becomes increasingly greater, as the applied current speed increases to any particular electromagnetic coil, as in the EMMA craft. As the electrons approach the speed of light, time is distorted, while a magnetic field is formed around these speeding particles, which tends to correct the time distortion. So, it is the magnetic electrons that act as a restrictive barrier at right angles to the electron current flow, preventing the electron current in the conductor from reaching or surpassing the threshold of light speed, which is approximately 186,000 miles per second in a vacuum.

Then, the key that causes this change in the speed and orbit of the subatomic particles is the alteration of the normal transit speed of the surrounding magnetic field, because once this is accomplished, the magnetic restrictive force or barrier is lifted. Then, the accelerated magnetic electrons need to be integrated or properly combined with the current flow, so they can act to boost the normal Earth current flow speed. When this happens, the system becomes a positive type assembly of attraction or high energy attractive field, similar to a gravitational field. In the case of the EMMA craft, its design features cause the current speed to be boosted or reduced in speed, causing a change in the time continuum and atomic solidity. A more detailed explanation of how this occurs will be presented in Chapter Seventeen. Note that magnetic fields know no rate of speed, only neutralization.

So, as the electrons are sped up in their conductive path, they merge with the moving magnetic vortex. This is because they have escaped their atomic mass and combine with the magnetic field to form a moving vortex that changes the electron speed of surrounding atoms, similar to when an object is accelerated in flight and time is reduced. In other words, the combined kinetic forces of the electrons and magnetic field, affects all the surrounding matter to the point where the electrons and nucleus particles form more outward orbital paths, as their speed increases. Ejected electron plasma clouds that are merged with a craft's surrounding magnetic field, sets up an atypical, spheroid magnetic field, as well as a spherical electric field.

These fields can then be expanded and contracted as metered waves of varying lengths are passed through them. This causes the minute plasma clouds to constantly rearrange themselves giving off both photons and lightrons, as the plasma clouds collide. Lightrons are photons opposite that cannot be seen. As previously mentioned, matter and antimatter electron plasma clouds emit mutual quantum spurts of energy (photons and lightrons), that appear as both a quantum (particle) and a wave.

The photographs taken of the experiment conducted on January 18, 1996 (Chapter One), proves that when electrons are accelerated beyond their normal rate of speed in a conductor, expansion of matter energy, the wood for example, in the surrounding environment takes place, because a different field or new type of form field, other than the normal expanding magnetic field, was created. As expansion of matter energy occurs, the time continuum must also change its rate.

Therefore, by increasing the electrons speed in a conductor, time slows down and the force of gravity is also reduced. By slowing down the electrons, time and gravity are increased. It may appear that by increasing time, it takes longer to get to one's destination, but the opposite is true since more time is produced, gravity is multiplied and the arrival to one's destination is quicker. So, as time comes to a standstill, gravity has no effect, but beyond this, time reverses and so does gravity, as it becomes an outward rather than an inward force.

As the electrons approach the speed of light in a conductor, they change the coherent field within the surrounding vicinity of the coils. Here lies the key to asymmetric field changing, so as to produce field dependent propulsion or converting mass to a non-mass as explained earlier. Asymmetric propulsion is what the EMMA craft accomplishes through a pulse varying application, applying many direct current pulses to each coil, as the magnetic masses rotate in the fields, because at some point, when the permanent current is applied to selected coils, the asymmetric field occurs causing a directional change, which in turn causes a change in the current flow speed or time. It is important to evaluate how these electrical fields and magnetic fields can change the gravitational field symmetry.

7

Controlling the Lock Force

As the electron orbit speed changes, the particles within the nucleus expand and contract with them so that the entire atom structure changes. It is possible to accelerate the particles within the nucleus of atoms beyond their normal shells to the point of becoming a non-mass, which can be called the photon energy environment.

When one achieves a no mass or non-mass field condition, they have changed matter into energy to cause it to be exited from our particular condensed field of energy to an in between energy-field condition, where the former condition no longer has an effect on it. This in between condition or neutral zone, which exists between matter and antimatter universe's, now allows for a choice of application by the crew to induce a condition, that will cause the craft entry into any desired energy-field bandwidth or coordinate state of atomic solidity in the universes.

These types of changes can be seen in the photographs of this book to a lesser extent. When the moving magnetic field becomes saturated with charged particles, it places a charge on all matter atoms it encounters or passes through, so that the neutral matter atoms become electrified and can be drawn into or swept along with the moving electrified magnetic field. This created field, then becomes the dominant force over all matter, gravity and time.

Photo Pages 34 and 35 were taken of an experiment conducted on July 10, 1989. These photos depict a space time distortion, along with a plasma ejection and emanations, due to a change in the current flow speed or time as we know it. This is because the strong nuclear force or "lock force state of existence" is being changed and energy in the form of plasma is released, due to differences in matter energy consistencies between different locations in space. This allows one to observe or enter a different dimensional field of existence, such as a different planetary environment.

By altering the condensation frequency of the plasma state of matter energy in the same dimension, a space craft will release some plasma energy to make small changes in position, in the same planetary environment. We have succeeded in splitting the atom as a means of releasing its vast energy. Now it is time to develop a means to alter the matter energy's released energy, to change this energy into another strong force or lock force energy. When this occurs, the matter can then be locked in to its pre-programmed destination, as it vibrates like the entered dimension (planetary environment) and becomes part of it. We need to harness the energy of the atom to reach the stars. However, this involves a whole new process where small incremental adjustments to the nuclear lock force occurs. When this lock force is changed, the change is applied equally to whatever matter it is being applied to. When the correct energy consistency field is produced that matches an entrance coordinate or location, the craft and crew are condensed into a different state of atomic solidity upon entry, as the applied force field is being gradually reduced.

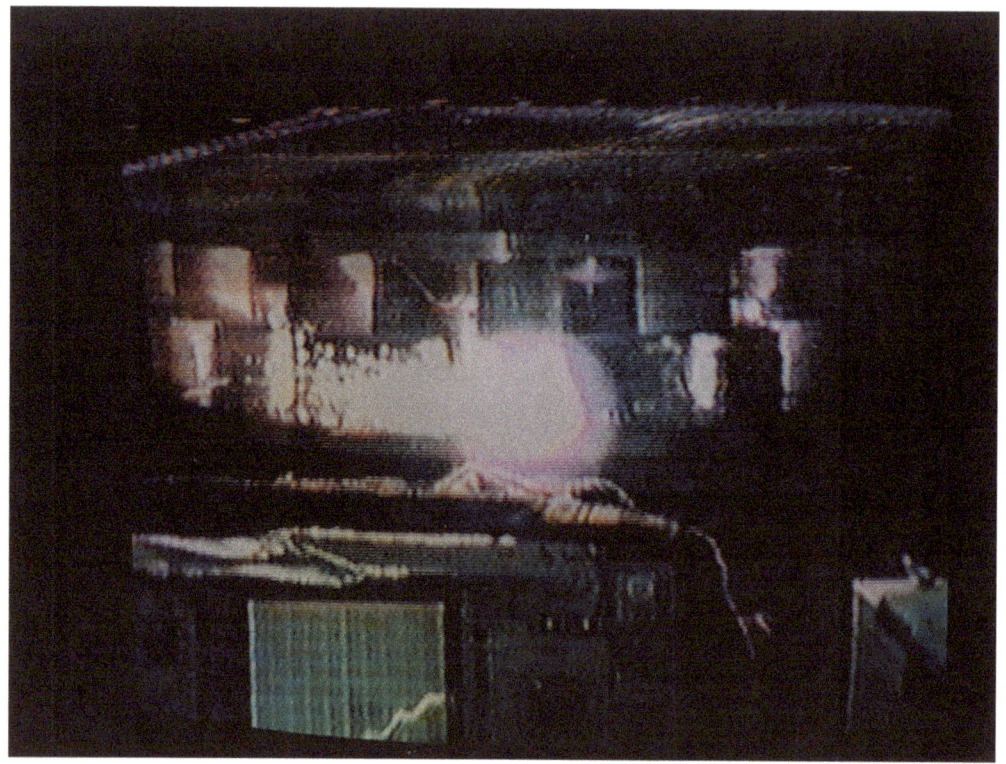

Since space is curved, different time entrances can be achieved into whatever space time zone one desires. By continuously compressing or expanding the applied field and then locking it in, different worlds can be entered that have a different atomic consistency for their atoms. Although the atomic mass of the craft and crew have changed into an energy form, the pulsed spheroid field keeps it confined within a certain vibrating atomic lock force or nuclear strong force range.

The lock force or strong force in the nuclei of all atoms is the vibrating field of whatever dimension you exist in. Change the lock force and you change your dimensional vibration, causing you to enter a different time or coordinate on Earth or in space. Otherwise, one remains in a certain locked-fixed shell condition, a prisoner of the Earth and its movements through space. Control the lock force and you can manipulate time as well as space. Picture all the particles within the nuclei of your atoms dancing to a certain tune. Change your tune and you change your location or address in the universes.

While it is the strong magnetic fields spherical configuration that allows the craft to continue along the time continuum, the time continuum is being altered. For example, expansion of the craft's mass into a confined energy form or no mass state, causes a slowdown of the time continuum. This in turn causes the craft's controls to reduce the energy release field. Since we cannot travel faster than light, by our universe standard, a different concept needs to be used if two world lines (different locations in space-time) have to be intersected.

Since the craft's mass is now total confined energy, the stress field or created gravitational field, diverts the craft's total energy, similar to light rays that are diverted in a strong gravitational field, such as around our Sun. The bending of the craft's energy, diverts the time continuum to seem to break through the light speed barrier by intersecting two world lines. Picture a long sheet of paper being folded in a curved manner so that opposite sides now touch each other and there is no longer any distance between them.

Space is part of the same flexible-pliable existence as matter energy, and just like the sheet of paper that can be folded over, a craft that needs to connect itself from one space location to another, can create a condition of space-time warping. Once entry to a distant location is achieved, space reverts to its original condition. A similar condition occurs as the wood snaps back or reverts to its original location, as seen and discussed in Chapter One. Perhaps everything has a "magnetic memory."

Since time is relative to existing conditions of gravitational light forces, light wave diversion by stress fields, constitutes the time essence (point of entry). So, by curving and intersecting the two-world line continua of light (like the folding over of the sheet of paper), which is the crafts mass converted into a confined energy form or light form, any time period or space time location can be entered, creating the appearance of faster than light speed travel, but it is only a one world line appearance that makes it seem so.

In other words, the energy form or craft and crew, is diverted into a curved path and around into another time continuum or space time location, with the assistance of the universes total power. Light waves appear to be straight lines of force, but this is only our one world line of appearance. Light waves bend and are intersected in any moment of arc with the proper diversion of the crafts-controlled force field. Is there mass? It all appears as mass in our one world line or four-dimensional space time existence, but is only a form of vortexing energy forces that by **mutual** attraction, appear solid and real to us.

This condition simulates within the affected mass, acceleration to the speed of light, as the craft remains stationary, so that time, gravity and mass are reduced to zero. Therefore, space becomes an instant entry, as the proper bilateral electrical, magnetic and quanta resonance rhythm is applied, as in a metered wave description, which alters the lock force. Each space time location can now be represented by a unique vibrational equation.

There is an equilibrium of this matter energy because as it expands, its energy time or energy matter makeup slows down after a certain threshold is reached. Picture a hoop spun forward given a reverse twist. At first the hoop travels forward increasing in speed. Then it begins to slow down after reaching a threshold coordinate and soon begins reversing. The same thing happens to changing of matter energy field resonance makeups. It is a condition of equilibrium that causes all light particles to be subject to the speed of light. This is why with the proper applied electrical, magnetic and quanta resonance rhythm, the consistency or elasticity of matter makeup fields can be changed.

As particles from our universe reach the speed of light, they are confronted with a stop barrier and are unable to enter the next adjacent universe. Likewise, as particles from the adjacent universe slow down to the speed of light, they are confronted with a stop barrier and are prevented from entering our universe. As previously stated, there has to exist a buffer zone barrier that if entered, one can receive particles so called from both universes, for changing entrance conditions to either universe.

If a craft from the opposite universe to ours, wishes to nullify the subatomic forces in its universe, the craft must transmit a particle focused beam containing particles from our universe. This is because their light speed particles or lightrons, are traveling far beyond light speed as we know it and they need to be slowed down for nullifying subatomic forces in their universe. The same would hold true in our universe. Faster than light speed particles must be employed, because their ultra-speeds cause all subatomic particles to be controlled and nullified to any degree.

The reason why we use the term lock force rather than strong force or nuclear glue force is because we want to stress it is this internal force in the nucleus of all matter, which locks all matter that is associated with Earth or any space body, into that particular body's movements through space. Unless one can alter their "lock force state of existence", they are going nowhere fast.

Super High Frequencies

Light, regardless of its intensity, travels at its known speed. This causes a standard or range of atomic solidity in our universe for all its matter makeup. In the lightron universe, the matter makeup standard is different, causing a different range of atomic solidity. Now, if particles can travel faster than light speed, this would indicate that all concepts of time collapse definitively. This is because they behave in exactly the opposite way to our normal particles. Instead of exhibiting infinite mass and with infinite energy, when they reach the speed of light in their universe, these particles lose mass and energy the faster they travel.

As previously mentioned, everything in the anti-dimensional world is reversed. The future is already occurring now. The effect can come first and then the cause. All one's memories would be that of the future. So, in a very real sense, on the other side of the stop barrier is our future. Time travel and thought travel are the same once time is eliminated, while our thoughts are a form of neutralization (that knows no speed), neutralizing an unbalanced condition, as we currently lack total knowledge and total awareness.

So, through the use of different intensity magnetic electron plasma fields that are expanding and contracting rhythmically, producing different resonating force fields, the time barrier is altered and the fixed field theory of unchanging atomic particle orbits that's unified changes. This is because the forces within the unified field interchange. Accelerated magnetic fields, along with electron ejection compounded, form a rotating expanding energy gradient appearing as a light form, that can cause an alteration of the time barrier. Time is disrupted and the time barrier (the constant speed of light) appears to change, as the applied force field forms a different spaced gradient of energy shells, so when matter expands, its subatomic particle orbits have expanded and its frequencies have changed to much higher frequencies.

Then space is mainly occupied by super high frequencies of energy of the cosmic ray spectrum and higher. Frequencies above cosmic rays on our spectrum scale would have such a high frequency, that they would appear motionless to us, yet coexist all around us, while we have no way of defining or detecting them. But, once we can produce frequencies above cosmic rays, a whole new science will become available to us. Endless energy—enough to create matter from energy—elimination of time decay (no aging or death), and the ability to enter different worlds by either increasing or decreasing frequency levels. The attraction of our bodies to earth is induced by the attraction of an invisible magnetic spectrum, which we have not been able to detect. Call it the gravity magnetic spectrum.

9
Manufacturing A 3-D Earth

We are living in a holographic universe where different energy consistencies abound. To manufacture a 3-D hologram Earth requires three types of laser beams. One laser emits photon energy, another magnetic energy and the third lightron energy. Lightron energy is the quantum energies emitted from the stars of our antimatter universe twin on the other side of the speed of light. By partially breaking down the magnetic barrier, a vortex of lightrons was produced around the device, thus producing a mixture of magnetic and gravitational fields (lightron energy), plus charged particle electrons or ions, which can affect the lock force, using certain pulse applications that expand and contract the force field. This is what occurred around the device at different times, where a time distortion appears and the coil assemblies begin to accelerate, as distance is being reduced.

It is an intermix of energy transference that occurs without disintegration, as opposing spectrums of energy and particles pass through magnetic neutralizer fields. It is an equalization that occurs, due to the loss of energy during a body's acceleration process and a changing of the time continuum. The greater the mass of a body, the greater it's speed, the greater its loss of energy and the greater the field attraction of the opposite dimensional energy. Each universe acts upon the other to cause the other to accelerate and rotate.

When we describe a transmitted hologram of light on Earth, we state that it is not real, because it cannot affect all our senses. We can see it and perhaps hear it speak, if voice transmissions are associated with it. Yet when we attempt to feel or touch it, we say it does not feel real. To us, there is no real substance to the hologram of light. This is because it does not contain the same energy consistency, as what we call solid matter atoms and molecules. However, if our particular state of hologram information (energy consistency for atoms and molecules), were stored in hologram computers, based on Fourier component technology, and then revised to that of a transmitted hologram of light, we could and would act in the same plane of existence, as the transmitted light hologram and it would appear solid and real to us.

As matter energy approaches the speed of light it reaches infinity and as it is somewhere in between or less than infinite, it occupies different states of energy matter quantum packets. Each packet is also of a certain extension, where the energy matter can attain different states of the dimensional field, such as solids, liquids, gases, plasma and various forms of radiation. In other words, each quantum packet is broad banded in its existing state to those who inhabit it, as they see exhibited a living hologram world around them and each living world being a unique quantum packet creation.

10

The Photon Lightron Chain Reaction Force

As previously stated, particles and atoms of each universe must be composed of three types of minute plasma shells, that represent three phase states of time. While one type rotates clockwise, another rotates counterclockwise and a third type does not spin, but acts as a neutralizing buffer zone force between the matter and antimatter rotating plasma shells or so-called particles.

One knows that if a neutron strikes other neutrons at a given energy force, a chain reaction occurs, until the matter makeup is completely fissioned (broken down) into energy. This is because this particular form of chain reaction, causes the electrons and positrons to change their orbits and strike each other and annihilation occurs. However, a way exists to cause these two particles to collide in a matter state that is far less dense, while the nucleus particles that need to be penetrated, can be penetrated by a much smaller energy mass, such as a photon. A high energy photon strike can penetrate the shell (outer surface) of a nucleus particle to affect the positron and electron, but it has to be done when these particles within the nucleus are in a dispersed condition.

Each atom is made up of different densities of these two subatomic forces and the closer they orbit each other at right angles, the denser the atom. One method for shell dispersion within the atomic nucleus has already been discussed and that involves producing a field that surrounds a craft. Now, we are speaking about a different process that involves concentrated beams of energy, which cause the dispersion of the atomic structure's internal energy, in a highly controllable, as well as safe fashion. Here, photons of light are employed to cause a chain reaction force of photons and lightrons, that are released in a manner that causes a modulation of the two into a concentrated beam. Although the energy of a photon is quite small, when combined with a lightron, the energy is vastly increased. With control of these two radiant energies in a waveform, a vast amount of energy can be created to propel any size craft using matter stress or change matter atomic consistency.

The photon chain reaction occurs when one gets monochromatic light, as all but one frequency is filtered out of the beam, leaving it weak and still incoherent. But, when the photon of a beam of light has the same frequency, phase and direction, it is coherent. Now, it is possible to make a coherent beam that does not spread out appreciably and if amplified, can be very intense. This same beam using the pulsed laser technique is the quantizing of the work function, where fo is called the threshold frequency and is the same as the cutoff frequency. Light, with a frequency less than that of the threshold frequency, will not have enough energy in its photons to free the electron or control the electrons orbital velocity around the nucleus.

This familiar amplification process was first predicted by Albert Einstein in 1917. Now, we need to take it a step further, so that we can control the electrons orbital velocity around the nucleus and obtain matter stress (gravity).

To obtain matter stress, where there is no time delay in absorbing a photon or quantum of energy, with a frequency greater than fo, the photoelectron requires a definite amount of energy all at once. Matter stress is accomplished when a coherent pulsed laser and variable ultrasonic sound is combined, that causes any isotope (atom) to vibrate, as a particle quantum beam and wave are formed and directed at any matter. This causes a gradual decay of the outer shell of the nucleus, along with the multiplication of photons and the release of electrons and positrons, in equal amounts, with the creation of matter stress by virtue of the photons and lightrons modulated, striking the orbiting electrons of the subject matter, changing their circular paths into elliptical paths.

It is a process that produces a chain reaction of photons and lightrons, as particle collisions of positrons and electrons occur in the surrounding environment, gradually releasing energy from the nucleus into the magnetic plasma vortex. The total mass, multiplied by the outer electron particles orbital acceleration, divided by the mass of the nucleus and its speed of rotation, determines the stability or threshold of an atom. In this process the atoms stability is maintained, as new particles (electrons and positrons) are created to replace the released particles, caused by the photon chain reaction force. The photon chain reaction acts similar to a blow torch where the outer surface of the nucleus is worn down. The energy released by this process can either be fielded, as in a plasma vortex or directed and focused to form a beam.

Direct current type pulses of changing amplitude sound waves are needed to cause atomic vibration, while the coherent pulsed laser will produce the matter stress, causing one-way motion. Pulsed sound waves, with changing amplitude, causes a varying atomic stability, making it easier for the photons to perform their needed function. If certain sounds affect certain matter makeup, its second, third and so on harmonic frequency will affect all matter makeup. However, if too high an amplitude is applied to the subject matter, complete disintegration will occur. A pulse laser ultrasound modulator beam must be constructed to produce this effect.

Once this method is achieved to cause matter stress and beam dispersion of the nucleus particles, then this penetration photon energy will be released in a chain reaction, similar to fission in a chain reaction of neutrons. However, the intensity of harmful rays that are emitted are far less dangerous and easily shielded, due to the fact that less density particle mass of protons and neutrons is being disintegrated, due to inner particle pre-dispersion. This device will completely revolutionize our current mode of travel and by reducing the concentration of the beam and focusing it on a flywheel, electricity can be generated. The craft will also produce electrical energy by tapping into the huge electrical and magnetic field, set up by the photon lightron chain reaction that takes place. This same process can be used to produce clean electrical power anywhere on Earth.

Mutual Force

By the proper modulation of photons and lightrons (picture a string of beads with alternating black and white beads), a carrier wave driving force is produced. Since negative and positive combined light waves are applied, the gravitational wave formed is invisible, as the two waves become one and their acceleration becomes neutral. Now, what occurs in nature is a reserve energy force (universal equivalence or universal gravitational force), that becomes activated by either negative or positive matter, thereby releasing its reserve force to whatever matter or antimatter is in the vicinity. The greater the mass (matter) or anti-mass (antimatter), the greater the reserve energy that becomes attracted or the greater the **mutual** attraction. As matter and antimatter attract, the gravity force is made to attract and it becomes an equal force that penetrates all matter and antimatter to a depth. Since all matter and antimatter contain a certain amount of each other's particles, the gravity force energy is quite evident in all dimensions or universes.

Its energy can be drawn upon by producing the proper particle wave carrier. The particle wave carrier will cause the reserve energy force to be attracted to it, while the reserve energy force becomes attracted to the particle wave carrier, so that a mutual force is formed. These are matched waves, where one is positive and the other negative, thus forming a neutral reserve energy force. The key to space entry is to be able to produce a particle wave carrier that is controllable, so that the reserve energy can be harnessed and controlled. With its discovery comes a great responsibility for its proper use. This means allowing for the normal functioning of the universes and a greater responsibility to preserve and protect the many living ecosystem worlds. We have a "poor" grade, when it comes to preserving and protecting our own living ecosystem.

Through the use of accelerated magnetic and electric fields, photons can be made to cross the barrier of light so that they modulate with lightrons, thus producing a photon lightron alternating chain, which can be called the **photon lightron chain reaction force**. It is this **type of photon** or **magnetic spectrum energy**, that can be used to penetrate from the exterior inward, the atomic nucleus of all matter, to cause a change in atomic solidity or cause matter stress, otherwise known as gravity.

To explain this photon chain reaction in terms of achieving a matter stress field around a craft, one must picture rotating beams of magnetic spectrum energy that can emit any pulse of a half wave, that instead of rising and falling, is pulsed like a straight beam of interrupted quantum. These are **spurts of magnetic spectrum energy** that are **emitted at different time intervals** to create acceleration of a craft and crew. The reason why we have never been able to achieve acceleration force from a magnetic spectrum emission, is because the waves are pulsed back and forth, causing them to be equalized. But, once the wave is rectified and sent out in straight spurts of interrupted quantum (like bursts of quantum bullets), the force is such that $F=ma$ (force equals mass times acceleration). Even though the mass (m) doing the acceleration is small, the acceleration (a), traveling at 186,000 miles per second is very high, so that the force (F) is great.

Light throughout all of space can be gathered, condensed and stored in cylinders to be used by **Dimensional Conveyors** as a propulsion system. This is because as light condenses, light produces the photon-lightron chain reaction force whose intensive strength draws other particles along its path to a collection center. Certain crystals can be doped with certain isotopes and activated by a coherent light beam to produce the gathering and condensing effect.

11

Straight Spurts of Interrupted Quantum

We know that gravity can penetrate all matter without any apparent adverse effects, while causing matter to accelerate. If we look at the high end of the electromagnetic spectrum, we see that high energy photons such as x-rays and gamma rays will pass through any matter to a degree. However, these high energy photons are harmful, as their short penetrating powerful waveforms disrupt and destroy living tissue. In addition, they do not cause matter to accelerate in the direction of their applied force. Therefore, we can rule out this type of quantum energy penetration, as a cause for gravity.

Perhaps what we are looking for is not a quantum energy force that travels in a waveform, but rather a straight time assimilated quantum of photon energy that vibrates into semi-quantum time delays, that has no waves, causing spasms of nuclear movement. Picture one half of a sign wave that is rectified to produce one half square wave. Now picture the removal of the rise and fall time so that what you have left is interrupted straight spurts of photon energy (see the following page). They would appear like dash lines on a piece of paper or bursts of bullets that have tremendous penetrating power, but without the harmful short penetrating waveform of x-rays or gamma rays. If one snaps a whip, the part closest to your hand has a larger wave. However, as the tip is approached, the wave is almost straight and due to its matter stress being concentrated toward the shorter wave, a snap sound is heard.

Now, these photon spurts or bullets only cause an orbital strain on the atom's electrons that orbit the nucleus, forcing them into an elliptical path, rather than forcing them to jump into a higher orbit or shell. Plus, instead of a waveform being produced, that acts like a Buzzsaw, where the applied energy is acceleration back and forth, these certain photon spectrum energies are applied in one-way quantum time assimilated spurts, very similar to a person gently pushing on a swing to increase the swings acceleration, at the prescribed time of the swings beginning point. As the electrons for example, travel in their elliptical paths around the nucleus, it causes the nucleus to be attracted to the furthest electron path and repelled on its closest path, creating a one-way stress in matter. These photon particles would act like bullets that stress the matter along in the direction of its applied penetrating rectified straight force. As these time assimilated quantum spurts are applied, neighboring atoms would be affected in the same manner, due to the photon lightron chain reaction that is produced.

So, to produce matter stress (gravity) where the photons are moving in a quantum (straight spurt) only, the square wave half pulse rectification system is acquired from a square or rectangular shaped periodic wave, which alternately assumes two fixed values for equal lengths of time, the transition time being negligible in comparison with the duration of each fixed value.

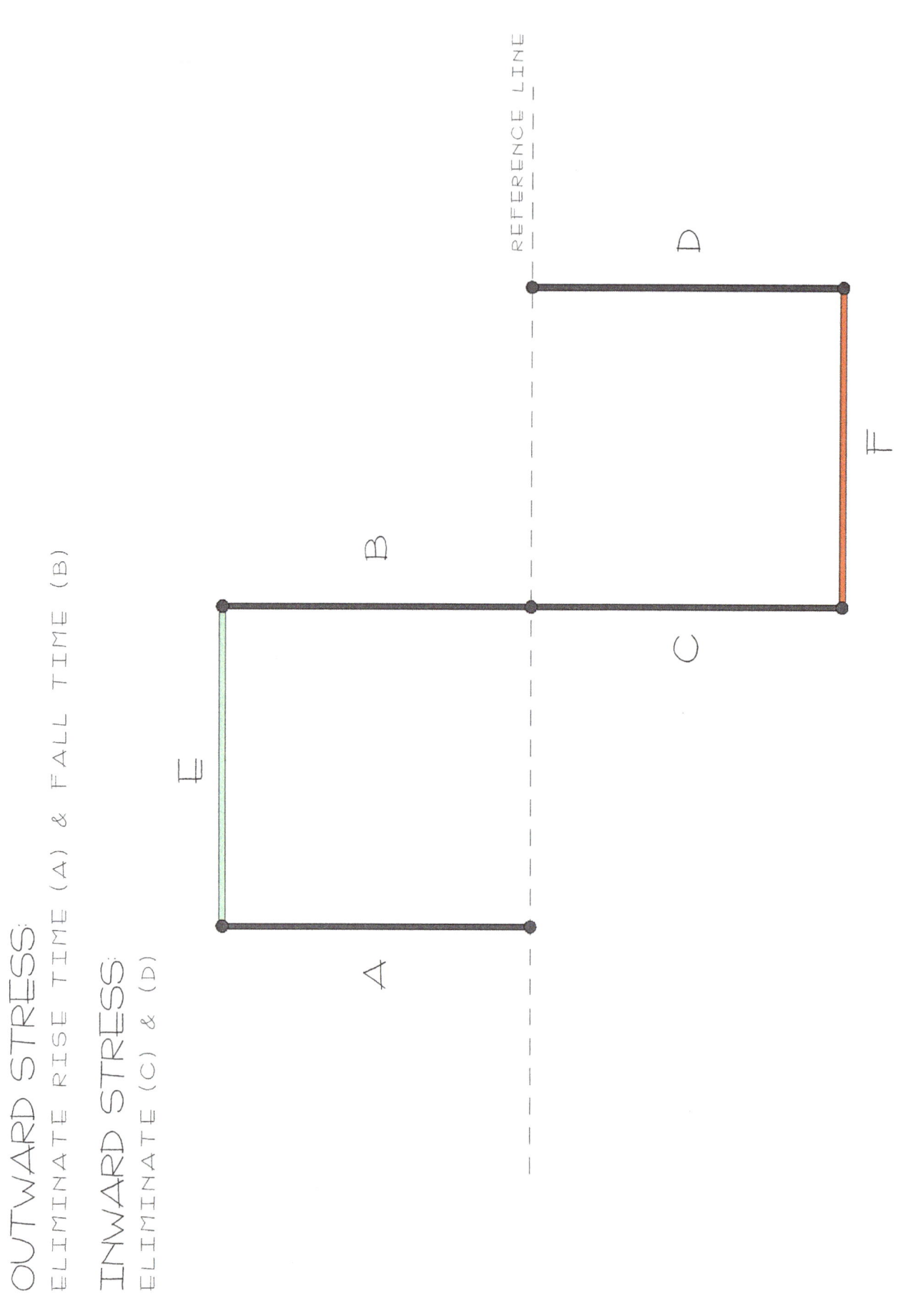

This type of wave rectification system can be achieved to produce a continuous range of electromagnetic radiation from the longest sound waves up to the shortest known cosmic rays. If one looks at the nomenclature of the square wave, eliminating the rise and fall time, one recognizes that amplification is only a matter of extension. The greater the amplification or intensity, the farther the force field or beam is extended inward or outward. This is because by producing matter stress (gravity), the dimensional atomic makeup field is either expanded or contracted, which allows the craft to enter whatever pre-selected atomic makeup field they desire and then make small adjustments for position changes, after entry is achieved. It is a system of matter environmental manipulation that can vary time, position or field of any existence in the universes.

As these photons are moving in straight spurts of interrupted quantum, they strike electrons and positrons within the nucleus, causing them to expand in one direction in the path of the photon quantum beam, forming an elliptical path. Now the photons by striking these particles, constitute a chain reaction of the matter and antimatter subatomic particles, where one photon affects both an electron and a positron and this multiplying effect causes a chain reaction that is equal to atomic fission and fusion combined. To expand the surrounding atoms and molecules, the intensity of the outward stress is made stronger than the inward stress (see illustration on previous page) and for compression, the negative spurt is greater in intensity. Only a small amount of energy is needed to apply the photon lightron chain reaction. Please keep in mind, the processes that have been presented requires experimentation. One process is one-way stress in matter and the other is changing the atomic solidity of matter.

The following 1929 quote is taken from Einstein a Life (Page 175) by Denis Brian:

"My relativity theory reduced to one formula all laws which govern space, time and gravitation. The purpose of my new work is to further this simplification, and particularly to reduce to one formula the explanation of the field of gravity and the field of electromagnetism. For this reason, I call it a contribution to 'a unified field theory.' Now, but only now, we know that the force which moves electrons in their ellipses about the nuclei of atoms is the same force which moves our earth in its annual course about the sun and is the same force which brings to us the rays of light and heat which makes life possible upon this planet."
– Albert Einstein

The quantum theory states that the emission of radiant energy is not continuous, and so it is because outside forces act upon it.

The gravitational effect we feel on Earth is the result of activating and attracting an energy from space that has no waves or frequency and acts as a cosmic equilibrium force or universal equivalence consisting of a dual spectrum force. To free ourselves from this space time energy attraction, we need to first develop around a specially designed craft, an electron or ion magnetic vortex and then expand and contract it. This expansion and contraction process will expand and contract the total atomic makeup of the craft and crew it encompasses in steps, as in a quantum. These step time pulses need to be tuned to whatever gravitational environment the craft is in or desires to enter.

Now, in the case of the magnetic plasma vortex, as it expands and contracts, the atoms it encompasses will also undergo a photon lightron chain reaction process, releasing energy from their nuclei. This energy will form a nuclear shield against all outside forces and can be called a no mass field. Just as there are different bandwidths of light spectrum, there will be different bandwidths of no mass conditions. So, a no mass condition does not necessarily mean a total energy condition. It can mean it is a bandwidth of no mass fields combined with frequencies of different consistencies of known matter energy. When a no mass field condition is produced, the surroundings become free from the universal equivalences restrictive controlling energy, that maintains an environmental order of the light threshold ("c"), and restricts the ability to change energy into usable matter, that can be condensed into any solid matter form.

Although light is almost total energy, it is not a no mass field. However, it can be used as a reference point of action for understanding this condition. Keep in mind, although the moving electrified magnetic field acts similar to gravity and affects all matter atoms, it is not a gravitational field, because it does not duplicate the gravitational fall rate of a planet, but it is a step toward producing a gravitational field.

12

Producing A Gravitational Field

Now, by applying certain pulse rates, the craft can be accelerated along to a particular planetary environment or it can be changed to enter a different position or time period in the environment it is currently in or some other planetary environment. The key is knowing the gravitational rate of fall for each environment one is in or wants to enter and then applying certain pulse rates in accordance with the speed of light. It's a matter of dividing the distance at which an object is traveling in a particular gravitational field, with each second's passage, into the speed of light to come up with the proper changing waveform application. The changing waveform is very similar to the changing waveform of the whip. **The waveform is not derived from the formula for free falling objects.**

As an example, since the Earth's gravity acts on all matter energy within its field to cause it to accelerate (fall) at 32 feet per second squared, we can come up with a similar applied force to duplicate or eliminate it, using a combined electrical and magnetic field in a waveform. First divide 32 feet into 186,000 miles (or 982,080,000 feet) per second, then do the same for 96 feet, then 192 feet, followed by 640 feet, 960 feet and 1,344 feet. The first pulse is sent at any potential amplitude, say one watt, while the second pulse is sent at three times the potential amplitude or three watts then the third pulse is sent at six watts or six times the applied potential amplitude, followed by 20 watts, 30 watts and 42 watts applied potential amplitude. In this manner, each succeeding wave gets amplified.

These waves are sent in short interrupted pulses (bursts), with delay times of 100 millionth of a second, 10 millionth of a second, 1 millionth of a second and so on. On Earth a primary frequency of thirty-two feet pulse in hertz is sent, then the power is increased for the next seconds frequency of ninety-six feet in hertz that is sent and so on continued up to six factors and then repeated. The time-energy point changes with each factor. With an electrical and magnetic pulse, the craft becomes part of the electrical and magnetic field whose energy multiplies with each step (spurt), because its constant speed is automatically turned into energy. The factor rate can also be reduced to say 6-5-4-3-2-1. As a particle (for example an electron) moves along its path, under the action of the force, its momentum changes continuously as does its quantum lengths. Each gravitational field is only a byproduct of the time continuum of energy matter one abides in.

So, once the input power is determined along with the initial wavelength, the input power can be calculated, along with the frequency adjustment for each subsequent harmonic. Note that in the case of the gravitational force, the quantum packets need to be sent **in progressive extended lengths of <u>interrupted straight spurts</u>** as to the gravitational energy points with each second's extension, and to ensure the energy field buildup, some type of energy increase is needed. The key is to design a triggering timing circuit that can be varied, so that each time the magnetic field collapses, it creates an antimagnetic field, if the correct time sequence is applied, creating a mixture of the two magnetic fields. The timing circuit can be made to increase or decrease steadily, changing not only the power inducement, but also the time dimension itself, by making small adjustments to delay times.

Not only does the time energy point change with each factor applied, this device would cause matter to absorb energy at a rate that duplicates the rate of fall of any object, in whatever gravitational environment one wishes to enter and then travel in. Duplicate a planet's gravitational fall rate (time continuum field of existence), around a properly designed craft and immediate entry occurs. One must learn how to surf or swing in rhythm with their expanding matter energy to be able to change or use gravitational waves.

More sequences can be applied if needed. An omni-directional pole antenna should be used, preferably at the top of the craft, in order to pulse in an area around the craft at whatever strength and distance desired. The direction of flight is achieved by the difference in the reflector portion of the antenna. An elliptical wave is sent in an expanded mode in the direction of flight and compressed in the opposite direction. Bottom and top reflectors are used for expanding and contracting the wave pulses, either up or down, for lift or descent. This works well for smaller craft.

Larger craft will require particle beams, pulsed in the same time sequence as has been described, plus a surrounding force field. Everything within the craft's force field acts as a laser spherical instrument, as the crew and crafts molecules and atoms vibrate in coherence with the applied pulse rate, causing the craft and crew members to enter the same dimension or applied coordinate atomic consistency change. It is important that the matter makeup threshold is maintained at all times with different speeds and time. In other words, the field maintains an equilibrium and it is this balance that allows for entrances to different dimensions of time and space. All matter energy exists in different states of equilibrium. If the threshold is exceeded, which is the rate of absorption and emission of energy, then the atomic makeup becomes unstable.

Gravity waves have not been discovered because they are changing in momentum and quantum packet lengths. As explained above, it becomes necessary to duplicate the gravitational waves at each momentum. The applied energy must be increased to continue the increased momentum, so that it becomes continuous, as gravitational forces are. To reduce the gravitational waves, just the opposite system is applied where at each momentum, the applied energy is reduced. With the gravitational wave, the quantum packets need to be sent in progressive extended lengths as to the gravitational energy points with each second's extension. This enables a craft to enter any time energy point, because time is only a contraction or expansion of energy matter makeup.

The idea is to quantum packet these waves so that they retain their structure for a time, not spreading out, producing bursts of quantum bullets that have tremendous penetrating power. To illustrate, picture water waves in an ocean. The waves start off with a small amplitude. Then increase in height until they reach an intense point and begin to reverse. Now, at the intense point, a great amount of energy is released as the wave collapses. Again, if one snaps a whip, the part closest to your hand has a larger wave. However, as the tip is approached, the wave is almost straight and due to its matter stress being concentrated toward the shorter wave, a snap sound is heard and energy is released. Here lies the key to quantum power and the absorption and release of all atomic energy makeup.

13

Dual Spectrums and Harmonic Amplification

Typically, science texts, when attempting to depict electromagnetic wave transmissions, will show an electrical wave projected vertically and a magnetic wave projected horizontally. However, this depiction fails to show the antimatter dimensional transmission that accompanies what was just described. In addition, it fails to show the nucleus particles that accompany such a waveform. If one pictures the matter electrical particle wave projected vertically and the matter magnetic field projected horizontally, in between these two at 45-degree angles, they need to insert an antimatter electrical particle wave and an antimatter magnetic wave. In addition, at each peak of such waves, matter and antimatter nucleus particles always accompany them, in such a manner that they surround the peak cycle in a circular fashion. See the following page for an illustration that depicts this wave description.

Without imposing matters antimatter reverse twin particle waves into the picture, it is impossible to explain all the effects of these particle waves. One must keep in mind that matter and antimatter waves exist and can be detected only in their own dimension. As this structure travels, the mixture also rotates. Rotational matter and antimatter electric and magnetic fields occur throughout space.

The universal equivalence is like a variable conductor. Electromagnetic waves can be transmitted for long distances because they partially disturb the equivalence. When the transmission ceases, balance is restored. However, these types of disturbances occur mostly in the matter spectrum, while the antimatter spectrum is very weakly disturbed. The antimatter spectrum has to be violently disturbed for the equivalence to release enormous amounts of energy to cause a gravitational gradient in the equivalence to be used by a space craft or for other purposes.

A ray of light projected in one direction must have an antimatter particle beam in the opposite direction, so that they constantly neutralize each other. One beam we can see and the other beam we cannot see or detect. Some experiments, referred to as Einstein's "spooky action at a distance" that have been conducted, prove that a light beam projects front and rear. So, if a certain condition can be imposed on one of the beams, a condition exists where the other will continually act upon the imposed condition and all kinds of conditions and changes in the universal equivalence can be induced in either the matter or antimatter world. This is because an unbalanced imposed condition is set up in one or the others quantum packets.

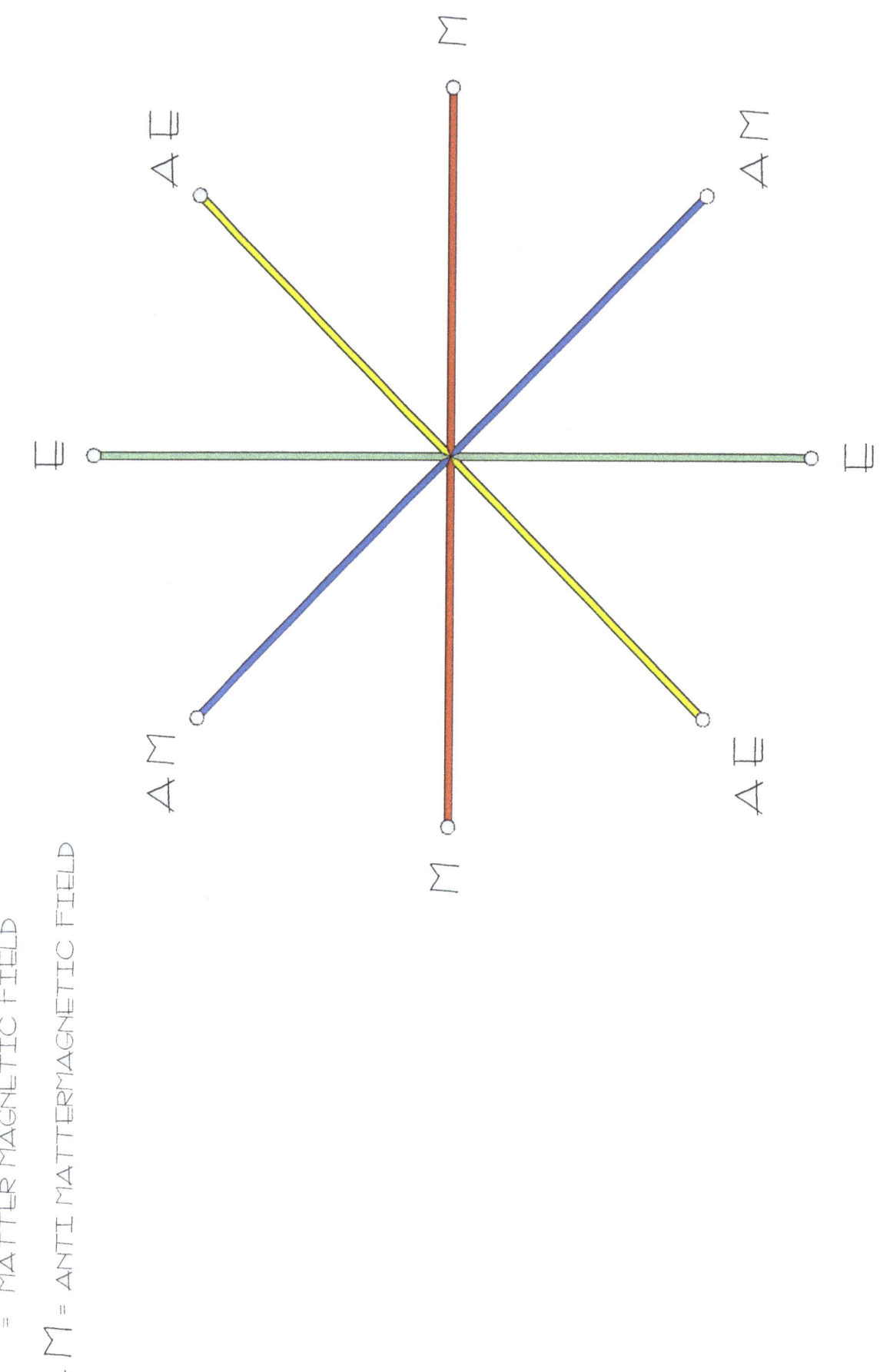

Matter and antimatter spectrums must somehow modulate to form a matter and antimatter balance. Unbalancing of either spectrum would cause the universal equivalence to act upon it, as in radio wave propagation or by exhibiting a force field energy release. To separate this balance or to change matter into antimatter spectrums, a series of sound and light pulses need to be produced omnidirectionally, so that its harmonics reaches into the antimatter spectrum by harmonic stages.

This is accomplished by harmonics pulsed in time-strengthened sequences as previously explained. So, as each pulse sequence is sent, it is continuously increased in amplification, up until say the sixth harmonic and then the sequence is repeated. Both photons and lightrons consist of micro mini-minite plasma clouds that rotate clockwise and counterclockwise. All matter is energy in a plasma field state.

A combination of different pulse frequencies, power input and delay time will cause havoc in the universal equivalence and cause it to release or produce vast amounts of energy. The time continuum is deterred and time stands still in between applied pulses. This is because we are dealing with an interrupted quantum of energy that vibrates into semi-quantum time delays that has no waves. Each time period on Earth had a certain quantum packet bandwidth and if one can duplicate it, they can enter that time period.

14

Dimensional Conveyors

If an electron is nothing more than an energy vortex of a certain charged consistency, then by encasing a craft with an electron magnetic vortex of different charged consistencies for its electrons, it is possible to duplicate and enter any electron charged consistency that inhabits the time continuum. In other words, as each electron vortex charged consistency inhabits different time coordinates in the time continuum, one only has to duplicate the makeup of electrons in whatever coordinate one wishes to enter. The craft acts as an electron energy vortex duplicator. This condition occurs due to the fact that neutralization is predominant throughout space in most cases. The craft will be attracted to the atomic structure of the environment of whatever coordinate one imposes on the craft.

As electrons are forced into another orbit when a catalyst current is sent through an electrically conductive wire, a similar condition occurs when an electrified magnetic vortex is created in any area. These freed electrons, as they establish new orbital paths or shells in the magnetic vortex, cause a change in the existing area makeup, which causes a change in the time continuum, with a release of energy (plasma). As different vortex intensities are applied, different plasma energy releases occur (see photo pages 34 & 35 as an example). These different applied intensities, along with the rotational speed of the vortex, causes new time continuum coordinates (locations in space). As this occurs, the vortex intensity applied and held in check causes the affected matter to enter its matched coordinate consistency. Now, since instant entry eliminates time, "c^2" is no longer a factor in the equation, so the equation becomes m=E or E=m, which means the conversion of either can be accomplished, where mass is converted to energy or energy is converted to mass. The conversion of energy to mass takes place upon entry to a matched coordinate consistency for atoms, as the force field intensity around a craft is gradually reduced.

Now, when the current is shut off in an electrical conductor, the electrons revert to their original coordinate consistency. Likewise, a very similar situation occurs when a magnetic electrified vortex is shut off as the craft and crew return to their original starting position or coordinate consistency. However, one must keep in mind, since the time continuum continues, a craft must change its original starting point, depending on the time continuum movement, so that it can re-enter at the proper point in time. You see, with this technology it is possible to return before you left. Look at it this way, if you left for six months to take a journey to other nearby solar systems, ideally, the best point in time to return to Earth is exactly six months to the day, after you left. If you returned to Earth on the same day you left, you would have no memories of your six months journey to other worlds. All your memories got erased, because you can't recall future events. It's like trying to recall a dream after waking up.

In the case of a space ship going to the Moon and returning weeks afterward, one recognizes the time continuum has changed and time has passed on. The crew upon their return will have to find out what has occurred during the time continuum change, while they were gone. As one can see, traveling throughout the universe, visiting numerous planetary systems in our galaxy, as well as other galaxies and then returning home to the time frame one left, requires a vast amount of knowledge with regard to different coordinate time continuum changes throughout our universe, as time proceeds at a different rate in each location you enter. Any civilization that can routinely accomplish this means of travel just described, would be so far ahead of our civilization, that comparing their scientific achievements and knowledge to ours, would be like comparing our scientific achievements and knowledge to that of a flock of sheep.

Rather than call this type of vehicle a space ship, a much more appropriate term would be **Dimensional Conveyor**, as the vehicle can be conveyed or transmitted from one point to another. It is a means of exiting one place and entering another, as time and distance are eliminated. It is a form of universal teleportation by the use of enclosed energy density vortex fields around an enclosed vehicle. The vehicle design needs to be able to hold as well as discharge the applied vortex field. A large conveyor design should ideally be spherical or a sphere that is somewhat flattened top and bottom, perhaps enhancing the stress application in those areas. The hull should have no sharp edges and must be able to retain both the magnetic field and the electron vortex energy field. A near perfect spherical field of both is needed around the crafts periphery. Then this vortexing spherical field can be shifted into any direction, causing acceleration in the direction of the applied applicant.

Magnetism is the neutralizing force between the electrical and the gravitational force that exhibits itself in the neutralizing dimension. Gravitational forces, as we know them, exhibit themselves in the antimatter dimension, but its force action is only felt in the matter dimension. Magnetism, on the other hand is felt in the matter and antimatter dimensions, but not seen by either. It is only seen in the neutralizing dimension.

Now, gravity, time and the constant speed of light can be altered by the alteration of the consistency of matter atoms and molecules. However, these changes must be pre-set and controlled by a crafts hologram computer with backup fail safe systems. When a certain magnetic gravitational accelerated field is pre-emitted for a particular destination, it matches the gravitational field of the destination and immediate entry is the result. The plasma field buildup around a craft is a force field that can be expanded or contracted by wavelengths. Adjusting the strong force plasma around the craft causes the craft's mass to either expand or contract its matter energy atomic consistency. The adjustment of wavelengths used has a direct relationship to consistency changes, by the use of long and short alternating sound waves.

15

Interstellar Communications

Since a magnetic field is nothing more than accelerated electrons in a more outward orbit, there appears to be no limit as to how far an orbital level they can achieve. Alternating, they can be received or transmitted by electronic equipment to a certain point in frequency. But suppose a direct current is maintained. There must be a circular type of vortex frequency, although our equipment is such that it cannot detect the higher vortex frequencies. This is because as the orbits of these magnetic electrons expand outward, they achieve greater acceleration that is beyond the speed of light. This is why they are invisible, but their force action on our material world can be felt. So, as electrons accelerate beyond their mass, they expand, become invisible and achieve orbital speeds having different light speed plus-factors with each increasing outward orbit. These types of "c"-plus constants (c= the speed of light in a vacuum), can be used for interstellar communication as well as a beacon to guide interstellar vehicles.

Each orbital speed of communication used would be determined by how far a transmission is needed. The further the transmission distance, the more outward orbital speed magnetic electrons used. For closer communications, where solar systems are only several light years away, the inner orbital speeds can be employed. Communication can occur between a **Dimensional Conveyor** and its home base by transceivers that are tuned to the coded vortex waves that the conveyor was pre-set at. As each circle of the vortex energy field expands, a different time loss energy is prevalent. It becomes imperative to have a transceiver that can be tuned to each sequential vortex coded circle.

Since the alternating wave magnetic systems now in use can only travel at the speed of light, they would not be appropriate for distant communications. The kinds of transmitters and receivers that are required would be able to segregate each vortex orbit of the magnetic electron speeds to determine the distance to send or receive for quick communication. Any message received from a civilization that still employs our present light speed technology would be of no value, because of the time involved to communicate back and forth and the lack of any worthwhile knowledge. On the other hand, if a receiver and transmitter were designed that can segregate each vortex of orbital electrons, then we can receive volumes of knowledge from advanced civilizations.

By inducing a modulation of voice and video to a correctly rotating vortex energy release of the transceiver, direct communication is possible. The transceiver will require a rotating amplified vortex tuner that can be adjusted, along with adjustable modulating voice and video transmissions. This will allow the light speed barrier of our transmission equipment to be altered, allowing for communication with a conveyor along its journey into a pre-set coded vortex circle.

Time and energy relate only to each coded vortex circle in the universes. Certain magnetic vortexes rotate in one direction to a point just beyond the speed of light and a little further outward, they begin to slow down, then reverse in direction. At this point, the time constant is reversed. Now, somewhere in between is a neutral magnetic vortex of magnetic forces. It is from this neutral point that a space conveyor can advance or regress in the time continuum simply by choosing and then initiating different magnetic vortex rings that are situated throughout space.

16

Superconducting Electrons and Positrons

Scientists have found that when passing a current of electrons through certain semi-conductor materials placed in liquid helium or nitrogen, the semi-conductor becomes a superconductor that can levitate a small permanent magnet. This action occurs because the extreme cold of these liquids forces an elliptical electron stress within the atoms of the semi-conductor, causing the permanent magnet to float, defying gravity. This is matter stress of a fixed nature, but if acceleration is to be achieved, a continuous one-way accelerated stress field has to be applied to all matter of an object or space craft and its occupants in order to achieve speeds of light squared ($E=c^2$), due to eliminated mass (m). When an electric current is passed through a superconductor, there is no resistance. When a one-way stress is placed on a material object, there is zero resistance in its path.

By developing materials that act as a superconductor of positrons, parallel wound coils can be developed with alternating layers of electron and positron superconducting materials, along with fiber optical UV layers in between each electron and positron layer. These types of electromagnets can also be placed onto a central pole within a craft and pulsed sequentially in the direction of travel or oppositely. In addition, focusable beams, previously discussed, can be installed in the hollow core of these superconducting electromagnets and rotated or sequentially pulsed around a crafts equator.

Now, when a current of electrons (forward time) and positrons (reverse time) are fed into these parallel windings in the same direction, all matter is now attracted to the one-way stress gravitational field of the coils, where the lines of force enter one side and exit the other side of an electromagnet. While the electron produced field affects certain elements, the positron produced field affects the rest.

In the case where alternation of the currents of electrons and positrons are applied, where pulses of one is more than the other, for example two positron pulses followed by one electron pulse, direct current dimensional differences will occur. This means that one is given greater intensity over the other so that time, mass and energy are altered. Positrons released from positron emitting isotopes and used in these parallel windings can create an antimagnetic field, while the electron current flow creates a magnetic field. By certain vibrations of like intensities of the two magnetic fields, stabilization occurs. By making one dominant over the other, interchanges between the two dimensions will occur. Different stabilization frequencies of the two can be applied to enter different so-called time periods and universal coordinates.

Matter has a certain vibration point and it is that particular vibration of whatever matter the positrons and electrons are pulsed at, which affects the atomic makeup and causes its separation into another level of vibration. As the positron pulses are more numerous than the electron pulses or vice versa, time (matter energy) can be stretched or compressed to whatever time coordinate one wishes to enter. This is necessary, because as one approaches the speed of light, matter would reach infinity, but by increasing the antimatter pulses, matter decreases in size.

Once the light barrier is broken, energy is released and the craft induces more matter pulse spurts compared to antimatter and increases in size. In this manner time is decreasing and increasing at the same time, but due to the effect of the craft's expansion and contraction it appears that the light barrier is broken, but it is really a time barrier that has been altered.

If one is a firm believer that the human mind can think of any idea, then that same mind can construct such a technology. One only has to evolve in one's thinking and soon that same mind will find a way to finalize the thought into a matter energy technology. The invisible thought becomes a visible resultant, by way of constructional application.

17

The EMMA Craft – Creating A Life Line to The Stars

The EMMA is a type of craft with a unique drive mechanism that will revolutionize all forms of transportation on Earth and in space. It is a means of non-polluting, quiet transportation that causes all affected atoms of the craft and crew to gain energy, rather than lose or expel energy, as is the case of a jet or rocket. These crafts convert stored energy due to rotational motion of magnetized masses, into linear acceleration. With the EMMA craft, when the threshold of injected energy of the atomic mass of the craft and crew is achieved, the overload energy is expelled as electrons, light and sound, rather than noxious gases. EMMA (Electro Magnetic Mass Accelerator) crafts employ what can be called a field propulsion system that requires no fuel. The energy injection principal requires energy to be pulsed intermittently (similar to pushing on a swing at the right moment), to cause linear vehicle acceleration.

The energy injection principal can also be compared to throwing a ball. Energy is induced from the person throwing the ball into the ball's atoms, causing the ball to accelerate in the direction of the applied action. The EMMA craft functions in a similar manner. It allows for properly timed energy injections, by the craft's programmable computers, so that a continuous linear acceleration is applied in any direction of travel. The EMMA craft can be programmed to accelerate at 32 feet per second squared or one-g force in the direction of flight, while traveling in the Earth's atmosphere or in space. While the vehicle starts off moving slowly, it quickly accelerates to several thousands of miles per hour toward its preprogrammed destination. But, at the halfway point to any destination, a reverse stress is applied.

Using the EMMA drives within the craft, the conservation of angular momentum energy is conserved up to a point. Then a magnetic restrictive force is imposed 180 degrees apart, by applying a direct current (versus a pulsed current application) to selected electromagnets (see the following pages that illustrate a 540-ft diameter Globe-Shaped Craft). This causes the large bullet-shaped magnetized **soft** iron masses, that are being accelerated through the hollow core of these pulsed electromagnets in opposite directions, to be temporarily slowed down. So, each time these magnetized masses hit their magnetic restriction, as they accelerate through sequentially pulsed electromagnets, their stored rotational energy gets released, producing a linear pulsed acceleration force, resulting in an inner atomic stress to all atoms of the craft and crew, in the direction of flight. This means that these hollow core, ring-shaped series of electromagnets, along with their magnetic masses that accelerate through them, function in stacked pairs, so that they act as duplex systems.

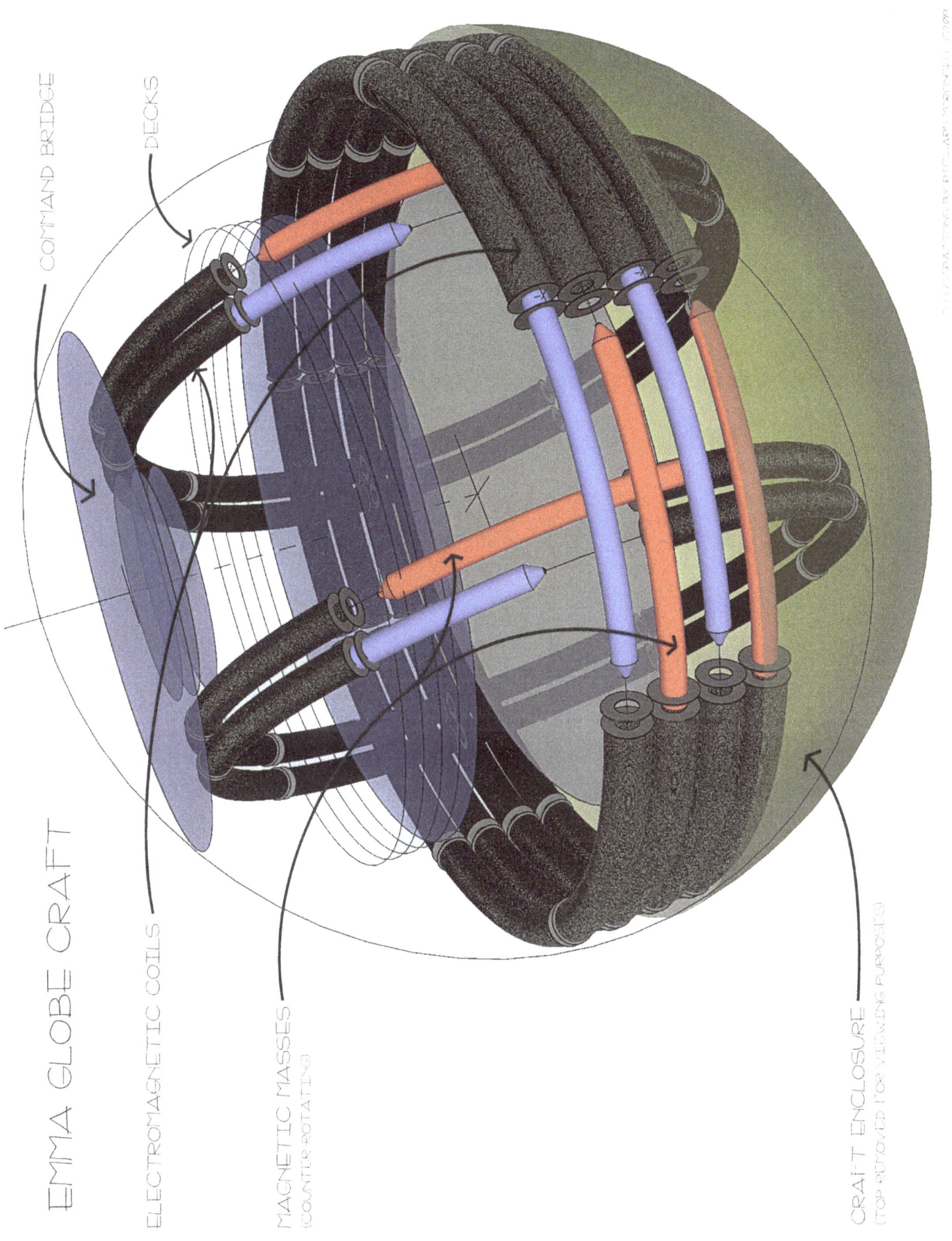

THE EMMA CRAFT – CREATING A LIFE LINE TO THE STARS

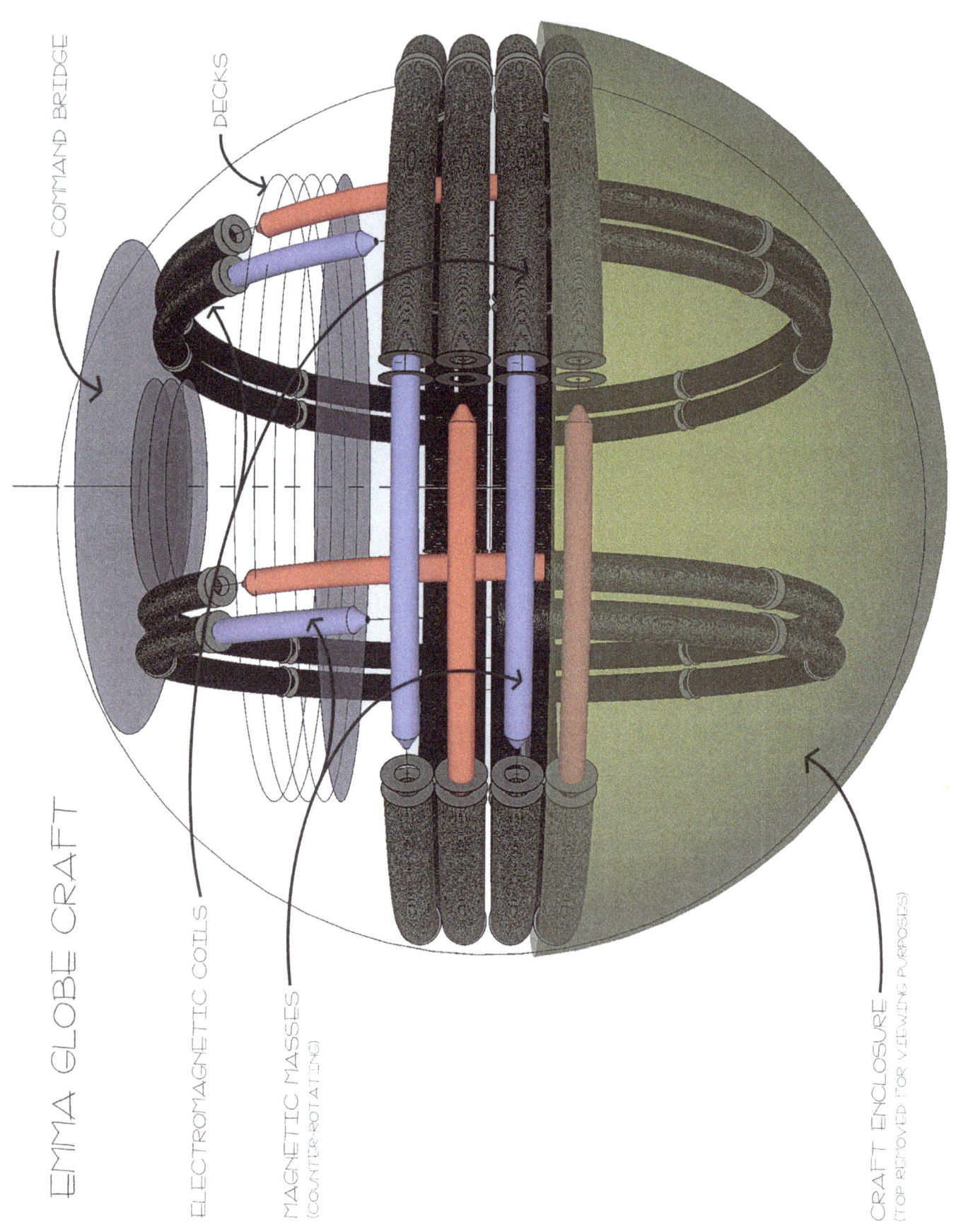

60 | TAPPING INTO THE TOTAL POWER OF THE UNIVERSE

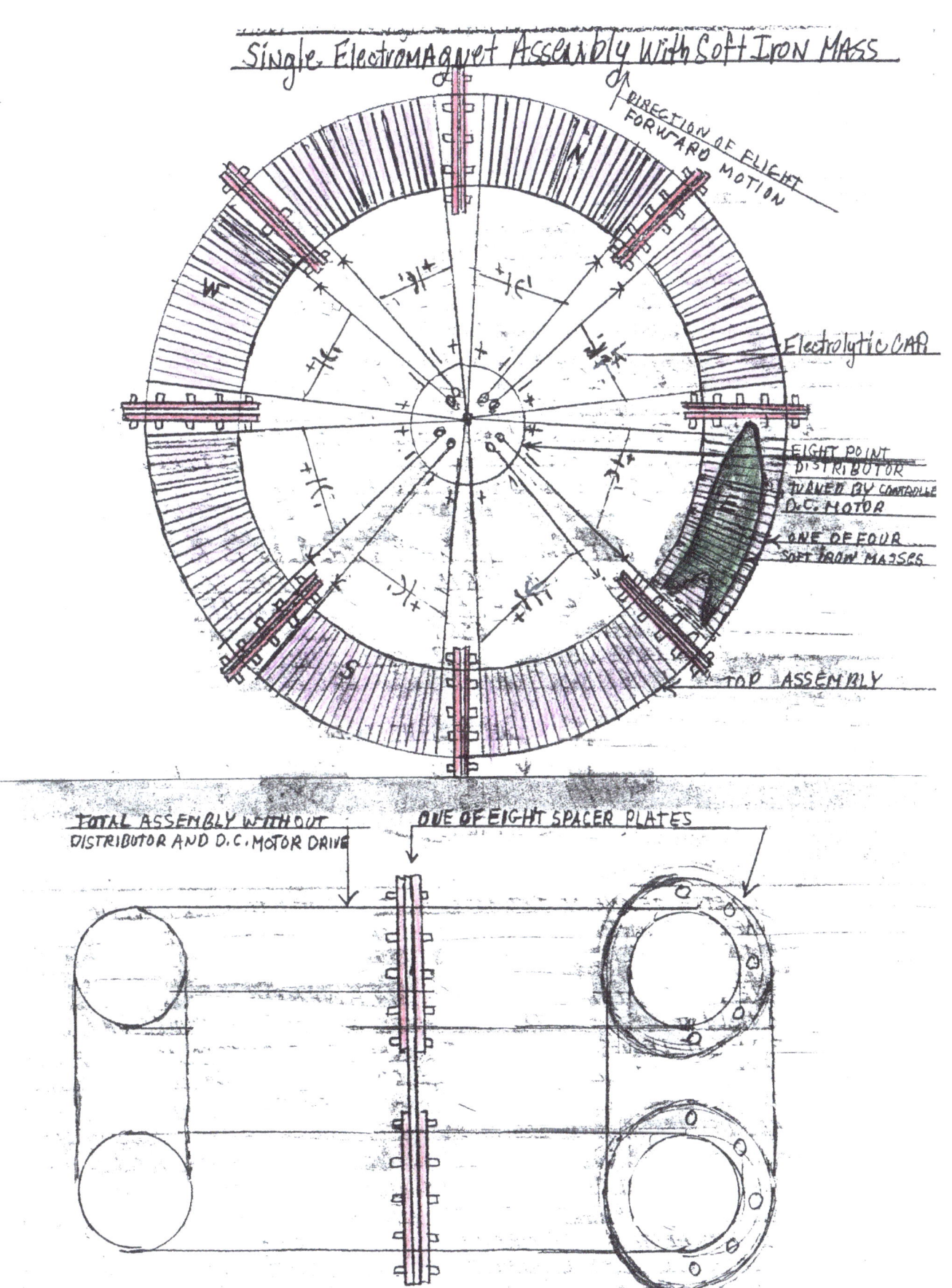

In order to better understand how stored energy created by rotational motion can be converted to produce a linear acceleration force, picture yourself standing in a large field. Suddenly, from behind you, someone runs up to you and thrusts both of their palms into each of your shoulders, at the same time, while applying equal force to each shoulder. This action causes you to be thrust forward in the direction of the applied force. Now, if that same individual travels with you and repeats this action over and over again, in intervals, you will continue to accelerate in the same linear direction. Similarly, each time the rotating magnetic masses traveling in opposite directions, hit their magnetic restrictions, 180 degrees apart from each other, a linear acceleration force is being produced, as the masses are traveling in the same direction, at that point in their counterrotational movement.

As the EMMA pulses these magnetic masses through each succeeding electromagnet, the masses act to either accelerate or decelerate the magnetic lines of force that surround each electromagnetic coil, above or below their "normal" transit speed, creating an effect on the time continuum, along with an inner atomic strain in the direction of flight, when the masses are restricted 180 degrees apart from each other. This is the principle behind the EMMA, where increased stored energy from the rotary motion of the masses, is converted to linear acceleration, as electrons and magnetic fields in and around the coil windings increase or decrease in speed, due to the use of magnetic masses that accelerate through sequentially pulsed electromagnets. Again, the whole concept is built around the control of the electron speed in a system of electromagnets within a craft that can alter the time continuum. Stopping the time continuum, stops the force of gravity and by certain adjustments, a combination time continuum and matter energy atomic consistency is produced, creating a "porthole of attraction."

But what is the mechanism within these duplex stacked EMMA drives that allows changes to matter, time and gravity to occur? The answer lies in the fact that any hindrance or restriction to a magnetic field allows such a field to give up or transmit energy. For example, as magnetic lines of force cut across an electrical conductor, energy is transmitted to the conductor, as these lines of force decelerate and a current of electricity is produced in the conductive wire. Surely, applied energy is induced to accelerate the magnetic lines of force in a set direction, but as they meet their restriction, energy is released in a decelerated state to conductive elements in our condensed matter energy dimension.

With the duplex EMMA drives (seen in the illustrations), we can control the acceleration or deceleration of the electromagnet's magnetic lines of force as the masses accelerate through their core, which causes the lines of force to release or gain energy. For example, If the magnetic masses have the same polarity as the electromagnets as they accelerate through them, the magnetic masses lines of force are decelerated, as they meet their restriction, causing the electromagnets magnetic field lines to accelerate, as energy is induced into them, due to deceleration of the mass's magnetic lines of force. In other words, the magnetic masses give up some of their magnetic energy to the electromagnets magnetic field, causing the electromagnets magnetic field to accelerate.

Now, if the magnetic masses are made to rotate in the opposite direction or the electromagnets polarity is reversed, the electromagnets magnetic field lines decelerate. These actions will cause the electromagnets electrons to decrease or increase in speed and what occurs is a change in the time continuum and gravity. What this means is that matter energy consistency of the craft and crews' atoms are also affected and changed, creating a "gateway" into other existences or planetary environments. In addition, "slip holes" can be created, allowing for a craft to make small changes from one place on Earth to another, while expelling plasma.

The EMMA craft coil design uses parallel layers of superconducting wires, where the current flows in a single direction through each coil layer versus the standard back and forth windings that allows the current to flow in two directions. Single direction current flow allows for changes in the current flow speed. Fiber optical layers that pass different frequencies of UV photons between each electrical current layer is required, while superconducting layers are highly recommended over non-superconducting layers.

In summary, these craft employ a pulsed stressed acceleration, with the use of pairs of magnetic masses. This will cause a linear one-way stress in all matter, when the stored energy is accelerated upon release, while time is altered by increasing or decreasing stored energy. By the control of the acceleration of these magnetic lines of force, we can gain or lose energy.

Across the equator of a spheroid or oblate spheroid craft (see Graphics on Pages 60 & 61), four masses rotate through their stacked tunnel-like electromagnetic coil assemblies. As previously explained, one mass in each pair rotates clockwise and the other rotates counter clockwise and each pair is magnetically restricted 180 degrees apart, in the same positions, say the 12 o'clock and 6 o'clock positions, for directional flight. The use of four large masses creates a directional power increase and adds to the stability of craft.

The horizontal assemblies are much larger than the vertical assemblies and of course the diameter of the craft will determine the size and the number of electromagnets, as well as the size of the masses that are being accelerated. While the illustrations depict the vertical assemblies extending from the command bridge down to the lower decks of the Globe Shaped Craft, these assemblies should not cross through the crafts equator. Several pairs of vertical assemblies can be placed above the depicted large horizontal assemblies, while an equal number of vertical assemblies are placed below the horizontal assemblies and spaced around and bolted to, the interior walls of the craft.

The magnetic lines of force are concentrated in each sequentially pulsed coil core, aiding in the mass's acceleration. Large carrier craft, say 1,800 feet in diameter or greater, can contain several smaller shuttle craft, that are kept secured to the lower interior portion of the craft, when not in use. These smaller craft can also be used to stabilize and help accelerate the larger craft and can be employed as a backup system in an emergency situation.

Also, in the case of the EMMA craft, a revolutionary means of anti-gravity takes effect, due to an electromagnetic screen that is created. The outward inertia is converted to forward motion or whatever motion is desired. This will mean that as the earth's gravity is overcome, the gravity-inertia of the rest of the universe is also overcome. An EMMA craft will be able to ignore the laws of inertia. The craft and crew will be enclosed in an envelope where neither gravity or inertia play any role.

For instance, when a force of whatever kind impels the craft and its crew members in a direction different from their line of movement, there is no tendency for their atoms and molecules to continue moving in their former direction, as the craft is surrounded by a revolving no mass field. The occupants can sit quietly without knowing that their craft actually was doing erratic maneuvers, such as right angle turns at extreme speeds and sudden stops.

Plus, the reason this craft will not burn up in the Earth's atmosphere, even when traveling at 20,000 mph or greater, is because if one considers a molecule bumping against other molecules in the atmosphere, subject to the laws of inertia, as everything else is, the molecules now find themselves in the gravity-inertia screen of the craft. Suddenly, these little molecules are entirely free. They no longer carry kinetic energy. They can bump into anything without causing the slightest friction. In other words, they enter the screen like a bullet, but strike the craft like a feather. As they exit the screen, the molecules kinetic energy is amplified as a result of the friction which was not possible within the screen and energy in the form of light is released.

So, if one can produce a force that can act like gravity and at the same time eliminate the Earth's gravity, they can accelerate any size craft, carrying thousands of crew members to achieve any speed, without any counter forces acting upon the craft. Take two objects. One weighing one pound and the other ten. Now, toss them into the air. It's much easier to throw the one-pound object as it requires less energy to do so. Now take the same two objects and drop them from a great height. Even though one object weighs ten times as much as the other, gravity has no problem accelerating them to the same degree and they both strike the ground at approximately the same time. It doesn't matter if one object weighs 200,000,000 pounds and another 200 pounds; they would be accelerated by the Earth's gravity force to the same degree.

However, in order to venture out into space, we need to understand gravity's total quality of action. For example, if an object rests on a table in front of you, it remains where it is largely because all the stars and nebulae of the cosmos are pulling on it and they are pulling on it in all directions. It is as if a million, million, million little wires were attached to the object symmetrically all around it and pulling it equally at the same time, in every conceivable direction. When you throw an object across the room, it goes in a straight line (aside from the Earth's gravity), because it is being pulled at every right angle to the direction of its flight, by the totality of matter in the universe, by all the stars and nebulae.

Thus, inertia in the familiar world is really gravitation, but not entirely the gravitation of the Earth alone or any single body near us, but the gravitational effect of every particle in the universe. It is the sum effect of a gigantic push, pull or field, depending on how you regard the still elusive gravitational mechanism. Note that one of the Buddhist sutras known as Avatamsaka Sutra considers "the entire cosmos as a nexus of conditions, in which everything simultaneously depends on and is depended on, by everything else." In other words, one recognizes the interdependence of all things.

18

Extracting Energy from the Earth

In 1999, a major modification to the experimental apparatus was made in an attempt to extract electrical energy from the Earth. This modification caused the production of a very strong neutral magnetic vortex around the counter rotating electromagnets. The photos in the following pages demonstrate that electrical energy can extracted from the ground. Approximately 200 bursts of energy took place over a period of 8 minutes, as the extracted plasma balls of minite plasma clouds, arose from the ground and collided with the charged particle magnetic vortex.

The small garage became filled with energy that was released, as collisions between "particles" took place, producing forms of "misty light plasma clouds", around the counter rotating magnetic fields of the coil assemblies. This extracted energy can be used to produce electrical power anywhere on Earth or gathered by a craft to be used for several different purposes. This includes a rapid intensification of the magnetic electrified vortex around the crafts hull, as well as tapping into the vortex for electrical power needs.

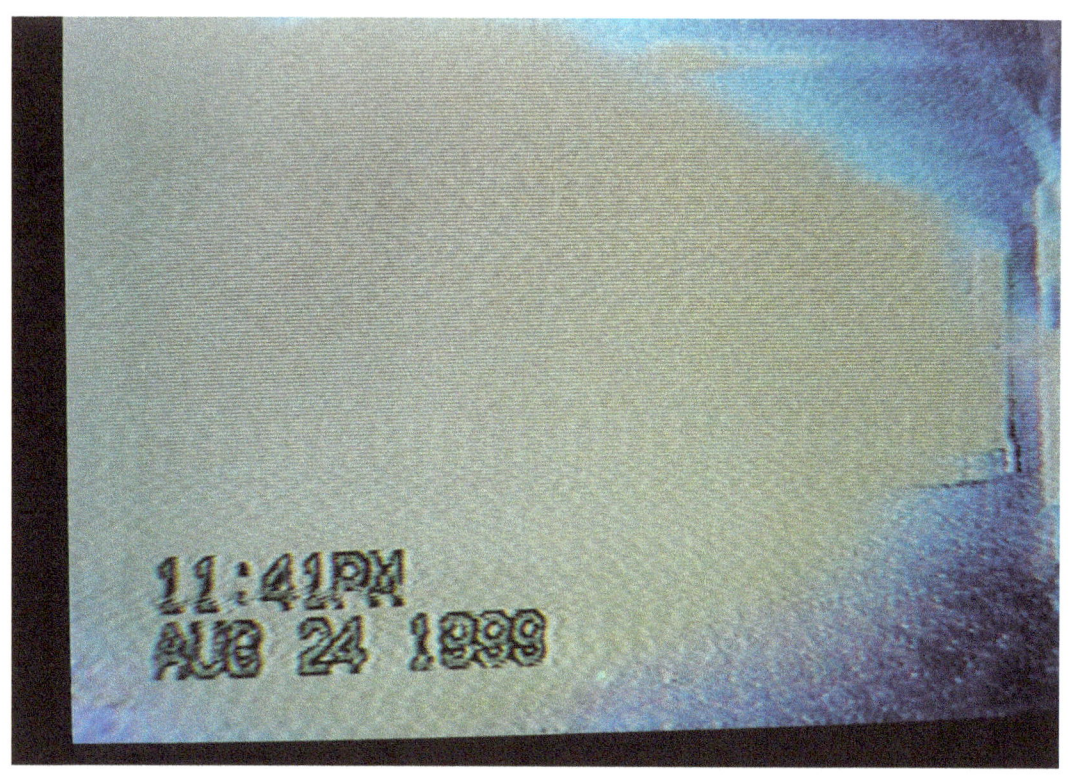

PART III

CONCLUSIONS AND FURTHER INSIGHT

19

Summary of Conclusions and Further Insight

Einstein showed that time is affected by gravity and that light can be bent to a strong gravitational field. Therefore, any change in time will affect gravity. If this is true, then the speed of light is no longer a barrier. This is because if gravity is altered, then what we call a standard second of elapsed time would also be changed. For example, if a second is elapsed in half the "normal" time on Earth, what happens to the speed of light that is measured in miles per second? Also, what happens to gravity whose fall rate is measured in Earth seconds squared, if a second of time has been reduced to half a second, by altering matter atomic solidity? Any change to gravity (by expanding or contracting matter), has to affect time and vice versa.

One can see that once we can control these factors and employ it technologically, the universe opens to our exploration. It makes sense that what we call time, is a constant change in the coded information of each atom, preventing us from seeing past or future atoms and events.

Matter exists because internally it is vibrating. Particle revolutions per second within atoms can represent a frequency vibration. Change this level of vibration, by the expansion or compression of matter and the matter makeup you exist in (your location) changes. In a coil of wire, the normal time continuum of electrons and atoms is occurring, but once a catalyst electrical current is introduced, acceleration variances occur, releasing energy in the form of what we call magnetic fields. Now to continue the increase in speed of electrons to produce a time continuum change, accelerated magnetic fields must be integrated with the electron current flow. Then, the speed of magnetic electron acceleration, combined with the speed of electron current flow, appears to surpass the speed of light, providing the magnetic field is employed as a booster, but it is really a time barrier or seconds that is being altered.

The magnetic field acts as a neutralizing buffer zone dimension between our matter dimension and an adjacent antimatter twin dimension or universe. You can call them a matter universe or dimension or an antimatter universe or dimension, but remember, each has a different constant for the speed of light. But it is this neutral separation that allows for the gravitational effect to occur in both universes or dimensions without annihilation. Each dimension feels the opposite dimensions force as a gravitational force.

Since the planets act as receivers, a variable transmitter type craft needs to be constructed to reach them. The EMMA craft design employs two means of transportation: gravity in the direction of flight (one-way stress in matter) and infinite velocity between two space time points. If humankind is ever going to travel to the stars, a much more advanced science and technology will have to be experimented with and properly developed. Rocket ships are never going to take us very far. A great deal of time and effort is being devoted to their development, but if personnel of principal evaluated the tremendous cost of placing just a few of our people on another planet in our solar system, they would see the folly of it all and devote their money, time and energy to a new type of space drive.

In the case of a rocket, a stress in matter is created in the opposite direction of the applied thrust of the rocket, while the crew members are being stressed in the applied direction of the energy release of the rocket. In other words, as the gases are expelled backward, so are the rocket crew members. In the case of a matter stress field craft, such as the EMMA, the craft accelerates the stress field (gravity) in the direction of travel, carrying along all matter in that same direction, while eliminating all counter forces on the crew and craft.

The craft's surrounding field involves electricity and magnetism, along with high and low frequency sound waves. As the masses revolve counter to each other, it causes the sound waves to compress and expand, due to the doppler effect.

As the electromagnets are sequentially pulsed, the craft also rotates when a greater restrictive force is applied to one of two selected electromagnets, in each pair of horizontal rings. This would be similar to a person approaching from behind you and thrusting their palms into your shoulders, while applying somewhat greater force to one shoulder. You will be thrust forward, but in addition, spin. It is this combined action, that causes a disturbance in the universal equivalence.

Now, a pulse type field that radiates outward in a spheroid manner will eliminate the gravitational cosmos equilibrium, as the antimatter world loses its attraction to the matter electric field and the crafts mass becomes a neutral no-mass field. The only power needed is the power to create the spheroid field and its adjusting intensity and pulse rate. The power for acceleration is provided by the universe pulling from all directions, because as the field is made stronger in any one direction, the craft will accelerate in the direction of the stronger side of the field. In other words, with proper distribution of the surrounding resonant energy, the space gravitational energy can be directed into any linear flight direction. The energies of the stars are the driving force and even though the craft is experiencing an explosive rate of acceleration (some 10,000 g's), no harm will come to the occupants or the most sensitive piece of equipment. This is because the accelerating force is applied with almost complete uniformity to every particle of the body or instruments on board.

Rotational effects by the spheroid craft, cause electrical particles to continually eject outward from its hull, thereby saturating all surrounding space and a non-gravity field or no-mass field is produced. Laser beams that are pulsed through the field become greatly intensified, because they have no equivalence (light speed) restriction. They can now be used to propel a craft much faster than the speed of light, as we know it. Science as it pertains to a non-gravity state, differs greatly to the scientific laws that apply in a gravity state of existence. The produced force field around the craft creates a nuclear shield against all matter and forces, so that gravity, meteors and atomic particles will have no effect on the craft or its crew.

There are two methods that are required for a **Dimensional Conveyor** to perform the functions that have been described in this book. One method suspends all molecular and atomic movement, freezing time and gravity, and the other method produces a form of gravity. Both actually function together, whereas in a 360-degree condition around a craft, certain areas in degrees around its periphery, time and gravity are not frozen, and a type of thrust is accomplished. This is where some testing will be required, using a small prototype model of the EMMA craft design, seen in the graphics of this book. At first, conduct tests using electron superconductors. Later on, positron superconductors can be developed and added.

Since the acceleration is increasing with time and time is decreasing with speed (Einstein's General Theory of Relativity), any destination can be reached within a reasonable time period. This is because the craft has near zero mass ($E=c^2$), so that the craft can travel at light speed squared. A small amount of mass is retained for the EMMA's permanent magnetic masses for directional flight. At the half way point to any predetermined destination, the craft's speed is automatically reduced by a retro applied matter stress to slow down the craft, as it approaches its objective. At the same time, a normal Earth gravitational field is injected into the crafts living and working compartments to allow crew members normal activities, as though they never left Earth's surface.

If a fourth dimensional object were to cut through our third dimension, only the part of the object that is actually in contact at any given moment with our reality, would be perceived by a third dimensional observer as physically solid, even to a point of resembling something that the observer can relate to, like a spaceship. However, the object itself could never be thoroughly understood for what it actually is because our senses and even our instruments are not designed to perceive entirely the true shape of anything, that exists beyond our range of comprehension.

Understandably, Einstein's General Relativity Theory plays a large role in the development of a new means of space and air travel, while quantum theory plays an equally important role. However, modifications to quantum theory are necessary, as this theory fails to take into consideration matter energy atomic solidity changes. The theory states that atoms do not change with the passage of time, but makes no attempt to explain why we can't see past or future events that are made up of these same atoms.

In addition, quantum theory claims that atoms of each kind, maintain the same fixed separation for energy levels or shells where particles reside, no matter where you are in the universe. This claim fails to consider acceleration differences throughout space that have an effect on atomic solidity. For example, as space time bodies travel faster, each atom's particles must travel faster and orbit further out from the center expending more energy. This is because there has to be a harmonious relationship between atoms and the celestial bodies they make up, as they accelerate through space together. All matter atoms and particles associated with a space body's movement through space is simply a part of the whole, that becomes affected by the whole body's acceleration through space.

Also, quantum theory fails to explain gravity's effect on matter. In Chapter Eleven: Straight Spurts of Interrupted Quantum, we discussed how gravity can penetrate all matter without any apparent adverse effects, while causing matter to accelerate. When we looked at the high end of the electromagnetic spectrum, we saw that high energy photons such as x-rays and gamma rays will pass through any matter to a degree.

However, we know these high energy photons are harmful, as their short penetrating powerful waveforms disrupt and destroy living tissue. We also know they do not cause matter to accelerate in the direction of their applied force. So, we ruled out this type of quantum energy penetration as a cause for gravity and determined what we are looking for is not a quantum energy force that travels in a waveform, but rather a straight time assimilated quantum of photon energy that vibrates into semi-quantum time delays that has no waves, causing spasms of nuclear movement. We concluded it is these photon straight spurts or bullets which only cause an orbital strain on the atom's electrons that orbit the nucleus, that forces them into an elliptical path, creating one-way stress in matter or gravity.

Now, with regard to electromagnetic transmissions, physics text books show the electrical wave projected vertically and a magnetic wave projected horizontally, but fail to show the antimatter dimensional transmission that accompanies them. In addition, it fails to show the nucleus particles that accompany such a waveform. If one pictures the matter electrical particle wave projected vertically and the matter magnetic field projected horizontally, in between these two at 45-degree angles, they need to insert an antimatter electrical particle wave and an antimatter magnetic wave. In addition, at each peak of such waves, matter and antimatter nucleus particles always accompany them, in such a manner that they surround the peak cycle in a circular fashion. Without imposing matters antimatter reverse twin particle waves into the picture, it is impossible to explain all the effects of these particle waves.

Based on our analysis of experimental results, we concluded that energy releases from matter in motion, must be what we call time. Therefore, it makes sense that space is filled with a form of highly expanded, undetectable reserve energy, which acts as a cosmic equilibrium force or universal equivalence, that has no waves or frequency and becomes activated and attracted to condensed matter distribution points, such as planets and stars, in the form of gravity equalization.

Since each space time body travels with different speeds and has a unique gravitational field (although some might be very similar), it makes sense that each space-time body's atoms have a unique meaning for the term solidity, due to differences in particle orbital speeds and orbital distances from the center point, so that what we call solid matter on Earth, would not be considered solid anywhere else. Based on our experimental results, it became obvious that the matter in our surroundings can be made to expand or contract. This expansion or contraction of matter atoms and molecules, causes time and gravity to be altered, due to different quantum packet energy releases from the affected matter. As time becomes altered in the immediate vicinity of the field coils, a window view can appear or a porthole of attraction can form, if the field intensity is increased. Holding the field intensity in check, allows for a fixed electron shell expansion or contraction, to the surrounding atomic structures.

We reasoned, that all matter energy must be a part of a single, elastic, pliable existence, that is expanding in a universal acceleration gradient and that each space coordinate or location can be represented by a changing unique vibrational equation, where change, the equations variable component, represents what we call time. Anyone reading this, should have enough information to test and prove the principles that have been related here, providing they have the required resources, including a proper facility and expertise.

Careful testing using low power is recommended. It's always best to take things slowly and to be cautious. Keep the field intensities low so that only a viewing window opens up. Later on, tests can be performed making small increases in field intensity, producing "slip holes." Remember, the purpose of these tests, using a small prototype of a properly designed EMMA craft, is to open a door that will give us the freedom to travel anywhere we desire. It will be a step by step process to reach the stars.

We are familiar with magnetic and photon energy, but lightron energy is of a different charged nature that is attracted to our matter energy to cause the condensing of matter in our universe, while photon energy is attracted to our antimatter twin adjacent universe and performs the same function. Magnetic energy is the neutralizer between the two that acts on a certain level, as previously discussed. The antimatter dimension has antimagnetic, lightron and photon energy, while the matter dimension has photon, lightron and magnetic energy. The magnetic energy oscillates back and forth acting as a mutual force between the two opposite energies. The magnetic force also acts as a bounding force around particles and maintains a certain consistency of matter energy for each so called "particle" in each dimension, which depends on the consistency or intensity of the magnetic bounding force.

As electrons flow through a coil of wire, a surrounding magnetic field is produced that acts as a barrier restrictor, preventing matter energy from reaching or exceeding the light speed barrier or time barrier. However, once we break the barrier of the magnetic neutralizing force, a vortex of lightrons occurs, producing gravitational forces and fields. As mentioned earlier in this book, continuous expanding or contracting of these magnetic fields are needed to break the threshold barrier of this neutralizing field, that acts as a watchman between dimensions. It is due to this continuous expansion and acceleration or contraction and deceleration of the electrified magnetic vortex that allows matter energy expansion or contraction.

This process will automatically change the matter energy that makes up any dimension and allows one to select which consistency of any particular atomic solidity they wish to enter. If matter energy is expanded for example, the lightron energy effect within its matter makeup is reduced. This is because the lightrons become less attracted to the expanded matter, so that gravity is reduced as well as time. When matter is condensed, gravity is increased. Photons and lightrons are responsible for gravity in each dimensional universe and one cannot exist without the other, while neither can exist in a condensed state of existence without the neutralizing magnetic force.

So, by expanding or contracting matter energy in any area, time is altered, along with gravity and distance. Affect any one of these factors and the others change accordingly. Also, if distance is shortened because time has been altered, speed increases. Just as the coil assemblies in our experiments began to accelerate on their own, because distance was shortened, the EMMA magnetizable masses should accelerate within a craft, as time and distance is being shortened, while gravity is being reduced. As the large magnetic masses in the EMMA craft accelerate through each sequentially pulsed superconducting electromagnet, they float through the center of each hollow core.

Our experiments demonstrate one must first electrify a magnetic field to cause everything it encounters to become electrified and with the magnetic fields' movement, matter stress (gravity) is formed, as all electrified matter in the field is swept along with its movement, as the integrated ejected electrons become the magnetic field lines' driving force. Now, once the magnetic vortex becomes saturated with charged particles, an atypical spherical magnetic field is formed around any properly designed spherical or oblate spheroid shaped vehicle, producing a single inner magnetic pole. This is because the vortexing charged particles in the magnetic field, complete the field configuration to form a monopole magnetic field. Then, this vortexing spherical field can be skewed into any linear direction, using the EMMA drive approach.

So, what is gravity? It is nothing more than matter stress. Larger planets generally have greater energy losses and would require more space time energy input to balance them and would exhibit a stronger gravitational field. In addition, all objects, regardless of what they were composed of, would fall at the same rate of acceleration, because they would be struck by gravity bullets being attracted to a planet, in proportion to their mass. As an example, lead atoms would receive or experience more incoming "bullet" strikes, then aluminum atoms.

In a sense, scientific analysis of gravity is correct when it is stated that the heavenly body's attraction of matter is the cause of gravity. However, without gravity wave attraction to matter planets and stars, there would be no gravitational effect. Without any energy attraction, how can an effect occur? If gravity waves attracted particles of matter, the particles would be elongated in shape. If particles of matter attracted each other, their shape would be randomly formed. However, if gravity waves and matter particles attracted each other, chances are a spherical type shape would be formed. This is due to the fact that all forces become almost equal in a duplicate attraction. Also, if gravity was an attractive force of matter, the less dense matter would be attracted at a greater speed, the same way a lighter substance will be attracted at a faster rate than a heavier one to a magnet.

However, if there exists a form of undetectable space time reserve energy, that is attracted to all space bodies within our universe at a standard rate for each body, then this would explain all body masses falling at the same rate of speed toward a planet or star. Planets and stars gravity affect each other in the same manner as two unlike poles of a magnet, where the magnets attract each other. However, it is the fields that attract and not the visible mass of the magnets. The reason one heavenly body attracts the other is because gravity waves are diverted from one to the other and are intensified in the area between them. This is very similar to magnetic lines of force which are concentrated in the center line of aligned magnets.

It seems the greater the concentration of matter, the greater the concentration of gravity, but does this help us to discover the secret of controlling matter? Take two balls of the same diameter, where one is denser and heavier and stores more energy and pick them up. As soon as you do this, you know which one has more stored energy because it weighs more. Now, if you place both balls on a seesaw, the seesaw will dip on the heavier balls side, but if you drop both balls from a great height, they both travel at the same released stored energy and rate of acceleration to the ground. Then, could it be that all matter in a vacuum fall at the same released energy called time energy expenditure?

So, all matter energy must receive its stored energy from its closest mass. Why? Because no matter how many times you drop the ball, it will continue to fall. It seems to never lose its stored energy. One may ask, how can such a ball travel without an energy loss? If it doesn't get its energy from the larger planet mass, where else can it get its energy from? Space? Most scientists believe that gravity is an attractive force between two masses, where the larger masses attract the smaller masses. For instance, the Earth attracts the Moon. However, in space, the space shuttle will not attract a small nearby pin, even though the shuttle is a much larger mass.

It appears that when a smaller, denser mass is close to a much larger mass like a planet, it receives stored energy from somewhere, to cause it to be attracted to the planet at the same fall rate, regardless of the density of the falling mass. There seems to be an equilibrium of distributed stored energy to any size density of the smaller masses. How did this come about? After all, if denser heavier smaller masses had a higher fall rate, wouldn't this disrupt the travel path of a planet, causing an imbalance? There has to be either an inward or outward field system for each celestial body that causes this equilibrium attraction to the larger, denser planet or celestial mass. If there wasn't, there would be no celestial order in our universe.

It is apparent that gravity waves effect all types of matter, which includes subatomic particles, as well as waves, although in the case of waves it is less apparent, perhaps due to their speed. We know that objects fall at a rate of approximately 32ft/sec./sec. near the Earth's surface. Does this give us a clue as to the type of waveform that needs to be produced in order to duplicate the Earth's gravity? Chapter Twelve explained how to produce the proper waveform. By pulsing such as wave, the applied wave can cause a multiplication of gravity waves, very similar to the law of bodies falling to Earth at about 32ft/sec./sec.

Now, by intensifying the waveform by adding charged particles, such as electrons, the effect becomes much greater. You saw what occurs when a moving magnetic field becomes electrified with the addition of charged particles in Chapter One. Everyone knows that in order to nullify or multiply an applied force, that force must be met with an equal force or greater force. Since gravity forces are of a weak nature, magnetic forces can be made to be much stronger and if properly applied, can negate or multiply gravity and cause matter to be stressed into whatever linear direction desired.

Since it is known that charged particles can be attracted to and engulfed in a magnetic field, it becomes evident that these particles along with a properly pulsed magnetic positive or negative directional field will double as a falling object mass. Although this type of mass doesn't have such a great mass density, the speed of the pulse and its intensity will be a better match than a heavy density mass with a lower speed. These pulses can be applied millions or even billions of times per second and this negates gravity.

With regard to gravitational forces, it is likely there are different stages of gravity and that different particles play a role in each stage. It appears that the magnetic field, which attracts only certain elements, is a partial gravitational field and that one needs to alter this partial gravitational field stage to form a total matter (inward force) and antimatter (outward force) gravitational field state, whose potential can be varied for different purposes. The gravitational force is made up of gravity and antigravity and the magnetic force is made up of magnetism and anti-magnetism, as both of these forces exhibit themselves in the matter and antimatter dimensional universes and are related.

It may be that gravity is the master force, which has different stages of potential, all of which is derived from the universal equivalence. Perhaps all matter energy and forces are nothing more than gravity in different stages of progress or what can best be described as, "the total evolution of atomic particle space and its movements." We understand the laws of gravity, but as far as understanding gravity's total scope of action, we have a long way to go. When the gravity magnetic spectrum is discovered, humankind will have created a major scientific achievement in our light magnetic spectrum world. However, if we continue to accept the current theory of gravity, its true nature and potential will never be discovered.

Could it be since a magnetic field only affects certain elements it lacks a mate which is also magnetically inclined? If two dissimilar particles, such as a positron and electron collide, the concentrated energy of each particle is expanded into a pure energy form. However, what occurs when two dissimilar magnetic fields (magnetic and antimagnetic) are mixed, that are already in an expanded energy form? Could this be the secret of producing condensed matter, as well as gravity from a pure energy state? Will it form matter or antimatter? Does it depend upon which of the two dissimilar magnetic fields is more intensified? What kind of in between dimension is formed when they are of equal intensity? Chapter 16 explained how to produce both magnetic and antimagnetic fields in a highly controllable fashion.

What is controlling this hologram environment we call the cosmos? Is it a preprogrammed condition by a higher intelligence or intelligent forms of energy that permeate the cosmos? If so, how can we alter the program to serve our own needs? Does the answer lie in the control of the mysterious magnetic field? First, we need to break the gravitational code containment field in order to understand the cosmos we are contained in. Then we need to set up a containment field of our own around a **Dimensional Conveyor** that blocks out or neutralizes this mysterious carrier force, which exists above cosmic rays and perhaps accelerates them.

If one can control the action of gravity, then control of an enclosed environment, such as a space craft, will allow any person to travel any distance, because time can be made to stand still or be increased or decreased to enter any time period or universal coordinate desired. Once such a traveling system is fully operational, increases beyond the speed of light are possible. Gravitational and magnetic fields appear to be designed to keep us prisoners of our Earth environment. These two forces appear to have been established as a means of creation and destruction, as well as a means of control. In order to reestablish our unlimited potential, we need to gain our freedom from these restrictive forces.

It is a combination of photons and lightrons that determine the existence of each dimension. The proper mixing of the two, must determine the entrance factor to whatever dimension or time period one wishes to acquire. Combine the two, photons and lightrons, and new spectrums will be achieved that can project entry into any time period or dimensional existence creating a lifeline to the stars. Alter their portioned intensities and you form a door to eternity and eternal life. Can the two, photons and lightrons be called the life force of the universes? It would appear so.

Perhaps all matter is a form of energy and all energy is in a state of living. We are not saying here that a stone is a form of life as we know it, but perhaps some other kind of intelligence on a different scale of living. For example, we have described a universe on the other side of the speed of light that is far more powerful than our own. Life form existence there would always be traveling at speeds greater than the speed of light (approximately 186,000 miles per second), as we know it. Therefore, if their life form energy could reduce their speed and frequency to that of our speed of light and enter the upper range of our matter world solidity in the visible light spectrum, we humans would appear to them as inert life forms, just like the motionless stone appears to be lifeless to us.

We can easily pick up a stone or any inert motionless object and move it anywhere we wish. Likewise, any energy intelligence that travels with speeds greater than our light speed constant, could do the same with us, as though we were an inert stone. Of course, they would have to have the means to enter our world or universe, while leaving their superspeed existence for a period of time. Nevertheless, if they could, we would be at their mercy. They would be gods or angels if they did good deeds and demons if they caused us harm.

As our universe expands and rotates in one direction, our adjacent twin universe rotates in the opposite direction. It would appear that our anti-universe twin is collapsing, but this may not be the case. It is likely that both universes are expanding and simply rotating in opposite directions. The universes should have a perfect rotational balance that acts as an accelerating, expanding neutralization. If there are a total of five mate pairs of universes or 10 universes, then all of them should be accelerating, expanding and rotating in pairs. It would appear that several of these other universes could be "octaves" above our more powerful antimatter twin. Then this could mean life form existence in these other "higher octave" universes, are even more powerful than life forms in our anti-universe twin.

Each of us has to recognize that what we see and feel as matter is made up of minites or miniature plasma clouds of a certain energy concentration and if your plasma cloud concentration suddenly becomes different, you can no longer perceive your normal surroundings with your five senses, unless you re-acquire the proper plasma cloud concentration. Also, plasma clouds of the past are a remnant energy. So, to go back into the past requires a conveyor or an individual to adopt a change in plasma shell concentration, to match that of the remnant matter energy plasma clouds of a particular past time one wishes to visit. What we feel and believe to be solid matter (solids, liquids and gases) is of a particular neutralized state of a holographic concentration of miniature plasma clouds that can be controlled for numerous purposes.

There is no such thing as time, only different concentrations of miniature plasma clouds or minites. Control these minite concentrations by the expansion or contraction of matter in any area, and you can perform any so-called miracle. Surround a craft with these minites and the plasma cloud minites can be vibrated in a way that the craft is surrounded by a magnetic and electric field condition, that is constantly changing within a metered wave, while photons and lightrons or mini-minites are being released and transferred between minite plasma clouds. This method of harmonic pulsing, using changing waveforms or changing quantum packet lengths, has already been explained. These metered waves or extended quantum packet lengths determine what dimension or coordinate the craft will enter or move to. Acceleration determines the consistency of all matter energy.

If mathematicians can determine the particular acceleration of each planet, sun and galaxy, then they can induce the same acceleration condition to a space conveyor via a magnetically enclosed plasma field, causing the **Dimensional Conveyor** to enter any particular planetary environment desired. So, once we can create certain enclosed energy fields that vibrate at certain imposed frequencies or certain vibrational accelerations along with certain consistencies, it will be found that we have tapped into the flow and total power of the universes universal equivalence. Only the initial imposed power to create such a field is needed, then the universal equivalence adjusts the space conveyor to the correct location within itself and immediate entry is the result.

Currently, the world's population doubles every 63 years. Population doubling is an example of a geometric progression. For instance, the current human global population is approximately 7.7

billion. In 63 years, it will be 15.4 billion and in 126 years the world's population will approach 31 billion people and after a third population doubling or 189 years from now, the population will be over 60 billion people. Then, after 10 population doublings, if that same trend were allowed to continue, the world's population will have reached 7.88 trillion people. Now, try to imagine, whatever town, city or village you currently live in, having one thousand times the number of people, homes or apartments and transportation vehicles.

But, once the population reaches the 23 billion mark in about 94 years, it is very doubtful we will have the planetary resources to sustain the majority of those people. We will have to triple everything in the economies of many nations so that people can have food, homes, jobs, clean water for drinking and bathing, fuel, health care, schools, hospitals, government services, transportation vehicles, roads and all that is needed for survival in a continuously expanding population. In addition, pollution of our atmosphere is altering our climate and reducing the potential for food production.

As you can see, the problem is that growth occurs so suddenly, that the population can become unmanageable without warning. In short, we are in danger of using up available space and required resources so suddenly, that we will have scarcely any warning to prepare for the onslaught of aggression and suffering that almost inevitably will result. Currently, we are witnessing increased suffering and aggression around the world as many families from poorer nations attempt to emigrate to wealthier nations. As the world population increases at an alarming rate, things will get worse. It may be overly pessimistic to predict that our technological society is doomed to an early end, but if we continue our present course, it is hard to imagine an alternative outcome.

Does expansion into space using our current scientific approach to space travel provide a solution? It has been seriously suggested that the Earth can relieve overpopulation by emigrating to other planets. However, even if we wanted to condemn our grandchildren to lives in air tight enclosures on the Moon and Mars, without any hope of playing in the fields or hiking in the woods, we would only extend our extinct time another 63 years, for the entire usable surface areas of the Moon and Mars combined is about the same as the land area of the Earth. In addition, to reach our nearest stellar neighbor in a reasonable period of time, carrying only a 50-pound payload, would require such a huge rocket, that its blast-off would incinerate the entire State of Florida.

On the other hand, by developing a new science that has been described here, we can produce unlimited clean energy on a planetary scale for a variety of purposes and in a variety of ways. We can create a new clean powered industrial revolution. We will be able to obtain needed mineral and metal resources from dead moons in our own solar system, using this new ultra-fast means of travel. With the production of controllable gravity beams, we can produce new, ultra-strong materials, by the compression of substances between gravity beams.

In addition, we can use controlled gravity beams to lift huge objects and transport them with ease around our planet and throughout our solar system. The control of gravity and time will open up a whole new era of technological discoveries, that will far surpass what has been written in the most imaginative science fiction novels. With time control technology, aging and disease can be regressed and one can achieve life over decay and death. Learn to travel to nearby solar systems in a reasonable period of time and then return to Earth.

It's time to wake up! It's time to open your eyes and look toward the star systems that were created for your enjoyment and exploration. If one can neutralize this force, we call gravity, one can free themselves from its controlling power and if one can create this gravitational force, doesn't this power to create gravity make one a creator of sorts? Since energy exists throughout the universe and all around us, all one has to do is use this gravitational force to condense energy into matter of any kind, by replicating any matter using templates, such as a loaf of bread.

In addition to opening up different corridors of science, we may someday witness firsthand, the process that created space itself and perhaps gain access to strange forces in our universe and other universes, along with gaining tremendous insight about ourselves and our creation. It may be possible to someday construct a specially designed craft, where each of the craft's numerous sections can simultaneously appear in different points in space and time. By walking through a passageway, one can enter another section of the craft and appear in another point in time or space, such as the past or different location in space, such as another planet in our universe or our adjacent universe. It would be similar to walking from one room into another room in your home and immediately finding yourself in a home or apartment you once lived in.

We can accomplish all of the above, by establishing positive goals. Warfare is destructive, while friendly competition between nations and people is healthy and constructive. While warfare has caused us to surge ahead technologically, now is the time for all nations to continue that advancement in a peaceful manner, enhancing the planets and people's welfare, while venturing out into space. Why not witness the beauty of creation, as we set out to look for other planetary homes, while protecting the only one we currently have? All of us have something to gain from this form of friendly competition. Leaders of each nation need to set their ego's aside and work together.

Each human being has been given two great powers. One is our emotions and the other is our curiosity. One must learn how to use these powers wisely. We have let our emotions take us down a path of endless warfare, death and destruction. Totally, these wars of destruction, represent the bulk of human history. It's time to let our curiosity take us on a different journey of peace and prosperity.

At present, thoughtful, rational, international planning to preserve our planet as a home for most of our race is missing. This needs to change. Currently, this planet is our only home, and we must do everything possible to protect it and to preserve our longevity. It's time for Earth to achieve a higher level of scientific and technological achievement, as well as a higher level of spiritual intellect, because both are needed to sustain a better way of life.

Einstein, as well as other scientists have argued that quantum theory is incomplete, while some scientists feel that general relativity theory may also be incomplete. Nevertheless, these are monumental works. However, changes are necessary to each of these great theories, so that we can take the next step, which will lead us on a path to discovering the secrets of inner dimensional freedom travel and clean unlimited power, along with enough energy to create matter from energy.

PART IV
GLOSSARY OF TERMS

Glossary of Terms

Atom- All matter in the everyday world of material objects is made of atoms. Atoms consist of a small, dense nucleus core, made of protons and neutrons, clumped at the center, around which orbits electrons. Examples of atoms are hydrogen, carbon and iron. The electrons can be described as cloudlike structures or shells that surround the nucleus core and revolve at high speeds.

Angular Velocity- When the angular velocity of a spinning wheel or disk doubles, its kinetic energy increases by a factor of four.

Antimatter- Matter consisting of antiparticles such as positrons and antiprotons that have opposite electrical charges, when compared to their matter "twin" particle. For example, while an electron has a negative charge, a positron has a positive charge. When particles of matter and antimatter collide, they form smaller "quantum vibrates". In other words, they are changed from one form to another. For each particle, there is a corresponding antiparticle. For each matter planet, star and galaxy, there should be a corresponding antimatter planet, star and galaxy. To understand why our scientists can't detect anti-matter galaxies, requires an anti-dimensional education.

Anti-Dimensional Education- As a magnetic field's movement causes an electrical current to flow in a properly positioned conductor, so does an electrical current flow cause a surrounding magnetic field to be produced. One can see there is a relationship between the two, but in order to understand this relationship, dimensional equivalence, along with inter-dimensional science needs to be studied. Therefore, one cannot study a condition that exists in this dimension, without studying conditions in an adjacent dimension. In other words, without an anti-dimensional education, one has acquired half an education, which equates to no real education. So, one of the many goals of this book, is to provide an anti-dimensional education foundation. Now use this knowledge, along with the expanded dimensional knowledge you have acquired, to go where no man or women on Earth has gone before.

Big Bang- The theory implies that all galaxies are racing away from one central point and that in the distant past they must therefore have all been extremely close together or compacted into an extraordinarily dense mass. It appears that the only way they could have reached their present positions and motions would be if some colossal explosion, generating energy beyond human comprehension, had sent them flying away from that central point. It has been found from observations made, that the faster a galaxy is moving, the farther away it is.

Conservation of Matter (Energy)- The principal states that matter can neither be created nor destroyed, but can only be changed from one form to another. Burn a log for example and energy in the form of light and heat is released, while ashes remain. The total amount of matter (energy) in the universe remains the same.

Doppler Effect- the apparent change in wavelength and frequency of sound and light due to relative motion between the source and observer. For example, as a plane is traveling toward you, the sound waves become compressed. As it moves away from you, the sound waves expand. In other words, the pitch is changing. Another example is when the train is approaching, the whistle has a higher pitch than when it is receding.

Dimensional Conveyor- A type of vehicle that controls matter energy atomic consistency, gravity and time. Dimensional conveyors are variable transmitters to the stars. Dimensional conveyors display a form of universal teleportation. The EMMA craft is a form of Dimensional Conveyor.

Einstein, Albert (1879-1955)- German-Swiss-American physicist (1921 Nobel Prize winner for his explanation of the photoelectric effect). Arguably the greatest scientist since Newton 300 years earlier, who had a difficult start in life. At first, he showed signs of mental retardation, not beginning to talk until he was three years old and not talking fluently until the age of nine. He hated school. Can you blame him? He got poor grades and as a teenager was expelled for being a "disruptive influence." Einstein flunked his first college entrance exam, passed on the second try, got mediocre grades and after graduation, could not find regular work.

In his spare time, Einstein worked on a hobby- devising a revolutionary new way of understanding the nature of matter, energy, time, space and gravity or what we can consider to be a new way of making sense of all existence. Beginning in 1905, Einstein wrote a series of papers that contained ideas. One of those papers led to the development of the laser, while another led to the development of photovoltaic power. Other papers gave us a new understanding of the universe we live in.

Faraday, Michael (1791- 1867)- English chemist and physicist. Faraday is often regarded as one of the world's greatest experimental scientists. He made many fundamental discoveries and interpretations that revealed magnetism and electricity to be two manifestations of the same physical force. He discovered that electricity could be used to make objects move (the principle underlying the electric motor), and that moving a magnet relative to a wire, would create electricity in the wire (The principle underlying the electric generator). Faraday had no formal education beyond grammar school.

Faraday's further work established that electricity and magnetism were alternative forms of the same fundamental kind of energy. This led to one of Faraday's strongest convictions: "that the forms under which the forces of matter are made manifest have one common origin." This was the beginning of what has become known as the unified field theory, which Einstein, in particular, pursued for many years, especially showing that gravity and electromagnetism have a common basis, a supposition that Faraday also held.

Force- This term is used in two related ways in science: it can refer to the fundamental interaction between two particles of matter that bind these particles together or propel them apart or refer to everyday phenomena that results from fundamental forces acting in concert. With regard to the latter description, let me give you two examples: The force of warm air rising can create winds that blow down trees. The force of muscles contracting can make balls hurtle through the air. In these two cases, the force is a mechanical agent. It is transferred from one body to another.

The energy consumed as a force may come from burning fuel or in a living organism burning food, or something else, such as an impact from another moving object. Once the force has acted on an object, it may transfer to that object, either of two forms of energy: kinetic energy or potential energy. Kinetic energy is energy in motion, while potential energy is energy of position or of being in a special position.

For example, an object on a shelf has potential energy corresponding to the force needed to lift it that high. As long as the force of gravity pulls on the object, it contains potential energy that may be converted to kinetic energy, if the shelf breaks.

Force field- is when two or more forces are aligned (flowing or traveling in the same direction), so that they act as a single unified field of force, like a flowing river. For example, magnetic fields and electric fields travel together, but typically at right angles to each other, restricting each other's force action. However, when they are properly integrated (no longer traveling at right angles to each other), they can have a great effect on the surrounding environment, as their force action is increased and as they rotate, they act as a no mass field condition, which can alter time and gravity. If such a vortex surpasses the speed of light at its periphery, time reverses and so does gravity, as it becomes an outward force or anti-gravity force.

Free Spirits- since they are free of matter energy and can produce it at will, they are free spirits who were once trapped or imprisoned (by the fallen gods) in a material body, but are now totally free. Once one reaches a level of spiritual energy being, they never die or suffer. When our physical body dies, our inner spirit body is released. But since we have been educated in the belief that the living body is the only form of life, we return to be reborn, only to suffer the perils and struggles of a material environment, trapped in a matter body. The spirit of human beings cannot be destroyed or created, since all spirits first existed. Once the spirit force energy is totally free from the body, a new concept of life will begin. The spiritual body or energy body is made up of matter from the neutral zone and is neither matter or antimatter.

Gravity- is the result of antimagnetic and magnetic field interactions, causing a reaction on all forms of matter energy. We see this in the strong force interaction within the nucleus of atoms, due to the movement of positrons and electrons. What needs to be done is to research and study gravity wave technology as it pertains to magnetic fields, meaning both kinds (magnetic and antimagnetic).

Infinite Speed- or infinite change allows for instant evolution and an intelligence too great to comprehend, because their condensed matter energy has been expanded throughout the entire universes, while removing all time fields for themselves. This allows these intelligent life forms of energy to control the entire universes and direct all events within. Entering a single matched condition (inner dimensional travel), between star systems for example, does not achieve instant evolution. Once one achieves infinite speed and instant evolution and permeates the entire universes, they have achieved all knowledge as to the life itself and the order of the universes, as well as achieving supreme rulership of the universes.

Inner Dimensional Travel- is a result of making inner changes to a craft and crews' atoms that can affect their location within the universe. When a craft and crews' atoms are expanded or compressed, immediate entry to a different state of matched atomic solidity occurs, regardless of distance. Different particle orbital speeds and orbital levels within and somewhat beyond the outer most shell of atoms, represents different locations throughout our universe. As within so without as above so below.

Laser- Light amplification by stimulated emission of radiation (Einstein 1917). A beam of light that is of a single frequency, hence one color only, that travels in one direction and most importantly contains coherent waves: that is, waves that move and fluctuate in steps, crest-to-crest and trough-to-trough.

Lock force- All matter in an appearance of lock force, radiates different intensity vibrational fields. To alter these vibrations, one must modulate a different frequency and receive a resultant to enter the lock force energy into another dimensional field of existence. This has to be accomplished while matter is in a plasma confined state or plasmoid form. The reason why we use the term lock force rather than strong force or nuclear glue force is because we want to stress it is this internal force in the nucleus of all matter, that locks all matter that is associated with Earth or any space body, into that particular body's movements through space. Unless we learn to alter our particular "lock force state of existence", we are going nowhere fast.

Mass- is a measure of the amount of matter in an object. It is the property of an object that gives it weight when gravity "pulls" on the object, according to current theory. On the surface of the Earth an objects mass is defined as being equal to its weight.

Matter stress- is gravity's effect on matter, where electron elliptical stresses cause matter objects to experience weight on the Earth's surface, or fall towards the Earth's surface. It is the control of the electron elliptical stresses that will cause all atomic nuclei to be stressed in any particular direction, due to application of the stress field. With matter stress, matter gains energy. For instance, an object dropped from a great height gains energy with time upon its acceleration towards the Earth, yet that same object attains the same mass during its acceleration, but gains "energy weight force", usually measured in terms of pounds. All objects on Earth have a certain amount of retained or reserve energy that is labeled as weight. Release the retained energy, by expanding (for example) or dispersing some matter, and the gravity carrier force is eliminated.

Optical fibers- Hair-thin fibers made from glass compounds through which light, usually in the form of laser pulses, is transmitted wholly contained inside the tube. A bundle of these fibers forms an optical cable for transmission of on-and-off laser signals.

Planck, Max (1858-1947)- German Physicist (1918 Nobel Prize winner for his work on quantum theory). Planck was the originator of quantum theory, one of the greatest scientific achievements. Here is a one of Planck's well-known quotes: "A new scientific truth does not triumph by convincing its opponents and making them see the light, but rather because its opponents eventually die, and a new generation grows up that is familiar with it." Throughout much of his life, Planck maintained a strong relationship with Albert Einstein—both as a mentor and professional colleague and as a valued friend.

Plasma- a plasma is an ionized gas, a gas into which sufficient energy is provided to free electrons from atoms or molecules and to allow both species, positive ions and negative electrons, to coexist. Tiny regions of plasma are created briefly inside electrical sparks, including lighting. Plasma is one of the four fundamental states of matter, which also include solids, liquids and gases.

Plasmoids- A plasmoid is a coherent structure of plasma and magnetic fields. Plasmoids have been proposed to explain natural phenomena such as ball lightning, magnetic bubbles in the magnetosphere, and objects in cometary tails and in the solar wind.

Principle of equivalence- holds that forces produced by gravity are in every way equivalent to forces produced by acceleration, so that it is theoretically impossible to distinguish between gravitational and acceleration forces, by any experiment (Einstein).

Relativity (Einstein)- Understanding relativity demands that we discard or suspend certain ways of thinking. Let's begin with: if two observers moving relative to each other, traveling at the speed of 100,000 miles per second, measured the velocity of the same ray of light, both observers would find the ray of light to be moving at 186,000 miles per second. In other words, the relative rate of motion between any observer and any ray of light is always the same, 186,000 miles per second. According to this theory, not only would the lengths in a line of a moving object be altered, but also time and mass. A clock in motion relative to an observer would seem to be slowed down and any material object would seem to increase in mass. Also, in space there is a non-existence of absolute rest in the entire universes. And two observers moving at the same velocity (same speed & direction) in time, may observe a space event at different time continuums.

This is because simultaneity does not exist for distant events. In other words, it is not possible to specify uniquely the time when an event happens, without reference to the place where it happens. Another way of expressing this is that the "distance" or "interval" between any two events can be accurately described by means of a combination of space and time, but not by either of these separately. This means that all events in the universe occur in the "space-time continuum." Three dimensions of space and one dimension for time.

In addition, every particle or object in the universes is described by a so-called world line, which describes its position in time and space. If two or more world lines intersect, an event or occurrence takes place. If the world line of a particle or object does not intersect any other world line, nothing has happened to it, and it is neither important nor meaningful to determine the location of the particle or object, at any given instant.

The world line is curved because of the curvature of the continuum in the neighborhood of the Earth. So, it is that every object attracts every other object in direct proportion to its mass is replaced by the relativistic hypothesis, that the continuum is curved in the neighborhood of massive objects in space. Simply stated, the world line of every object is a geodesic in the continuum. A geodesic is the shortest distance between two points, but in curved space, it is not generally a straight line. In the same way geodesics on the surface of our Earth are great circles, which are not straight lines on any ordinary map. Relativity also explains that the frequency of light in a strong gravitational field is less than the frequency of a similar light elsewhere.

Rotating motion- is a form of acceleration, because every moment a point on the rim of an aluminum disk for example, is undergoing a change in direction, which means that it's velocity (a combination of speed and direction) is undergoing a change.

Strong Force- is one of the fundamental forces in nature. It holds together the particles of the atom's nucleus, made up of neutrons and protons. The strong force or strong interaction, is about one hundred times as strong as electromagnetism.

Superconductor- The state of any of a handful of metals and alloys under an extreme condition of very low temperatures whereby it loses all resistance to the flow of electrical currents. Hence, these currents can flow without any loss of power. Recently discovered high-temperature superconductors are a new breed being made of ceramics, which are very poor conductors of electricity under normal conditions.

Supreme Being or Initial Grand Spirit- The Initial Grand Spirit created from itself uncountable numbers of spirits (minds), whose purpose was to create a matter base (universes) in the infinity and to increase awareness. Each mind can best be defined as a timeless and indestructible unit of awareness, each possessing a different viewpoint and each capable of infinite creation, because creation by a spiritual being is accomplished by the act of thought or imagination. By ones thought anything can be created, including universes that can be shared and experienced by all others. Every unit because it possesses a unique viewpoint is a source of its own infinity, as thought and imagination have no bounds.

Each human being is a unit of awareness and a part of the Supreme Being that became trapped in a material body. One may ask, what went wrong? How did so many spiritual beings each capable of infinite creation, wind up living on Earth thinking that they are only material beings?

Once the universe appears to operate on very simple building blocks and once those blocks are put into place and other arbitraries are introduced, the universe became extremely complex and solid looking. When this happens, spiritual beings become fixed or trapped in those universes and there are trillions of universes. In this state, spiritual beings have lost their power to change perspective in relation to the physical universe they inhabit. Perspective is what determines the size of a spiritual being. A spiritual being in an un-trapped state can change its perspective in relation to the entire physical universe, while spiritual beings on Earth are largely confined to the single perspective dictated by the physical bodies they animate, as they keep returning to Earth, only to suffer over and over again.

The act of repressing a spiritual being, entrapping it in matter, or otherwise seeking to reduce its vision, creativity or self-awareness as a spiritual being, is the act of trying to reduce a Supreme Being. If one reduces a Supreme Being's unit awareness (a spiritual being) by just one unit out of many trillions of units, one still reduced the Supreme Being's creation by that much. If the human spirit is released from the body, it can perform any action on any matter because it is the controller of matter energy creation. Once your human spirit is totally free, you will be free from bondage as you have been made to toil in labor and suffer in a matter energy form.

Time Dilation- Einstein's relativity theory says the faster an object moves, the slower time passes for that object, as observed by a witness moving at a slower speed. The amount of slowing, or time dilation, is noticeable only when objects move at an appreciable fraction of the speed of light.

The Human Brain- As the left side of the brain impulses its own messages, it creates a computerized-like memory and instruction bank for each individual's use. When one learns a certain skill, the memory bank robot takes over the body and performs those functions we call skills. One performs these skills without evaluating them. We have left the human robot in charge so to speak and are not aware of it. Now the right side of the brain, which if commanded by one who has control of his or her command center, could command many things such as heal one's body and the body of others, perform mathematical problems like a computer, read any persons mind, duplicate any future scene visually, construct mentally and physically any thought or idea, see other time coordinates and dimensions and last but not least, mentally enter any matter and feel its vibrations, all living as they are in a sea, because they are part of the whole.

One must learn how to relax, be calm and slow down the left side of the brain's functional vibration rate. Then one will enter the command center on the right side of the brain and perform many functions, which you now call miracles.

The Quantum of The Quality- A quantum is a discrete portion of energy of a definite amount, which by quantization pulses or spurts can allow an atom or molecule to emit or absorb energy. This is because an atom absorbs or emits energy in a series of steps and each step being the emission or absorption of an amount of energy that is called a quantum.

The energy in each quantum is directly proportional to the frequency or time assimilation of photon chain reaction spurts. The quantum of spurts quantity of magnification would always be an integral multiple of a definite unit. This unit can be called the quantum of the quality.

Now, if the basic step time of any specific matter is known, then that matter can be expanded or contracted by the emission of the photon chain reaction of quantization. They can be sent in integrals of the same multiple definite unit or they can be sent in an elongated multiple mode of varying spurts. By the multiple of each photon spurt, energy will increase with time and distance, in the same manner all objects on Earth fall at an increased rate of speed and energy, with time and distance.

In the case of nuclear reactors there is always a chance that the reactor may go into a critical stage if the cooling systems fail or the control rods become ineffective. But with the photon chain reactor, the fission occurs in a controlled manner because these waves are "chained" a little at a time. So, if the machinery becomes faulty, no chain reaction occurs. They, the combined waves, can be condensed or expanded to form controlled gravitational fields and can be used to release energy from the nucleus to form an enormous and powerful confined magnetic plasma field (vortex) around a space conveyor.

Universal Equivalence-or cosmic equilibrium force is a universal gravitational force.

Warrior- one that fights for the rights of all sentient beings.

www.ingramcontent.com/pod-product-compliance
Lightning Source LLC
Chambersburg PA
CBHW040543220526
45473CB00016B/3008